2018

INTERNATIONAL DESIGN EXCHANGE PROJECT

The 9th Hidden Space Project

RECUSED; Space & People

01

The Cass London Metropolitan University
The Sir John Cass Faculty of Art, Architecture and Design
Interior Design BA (Hons)
UK

02

Dongyang Mirae University
Dept. of Interior Environmental Design, in School of Living and Environmental Engineering
Rep. of KOREA

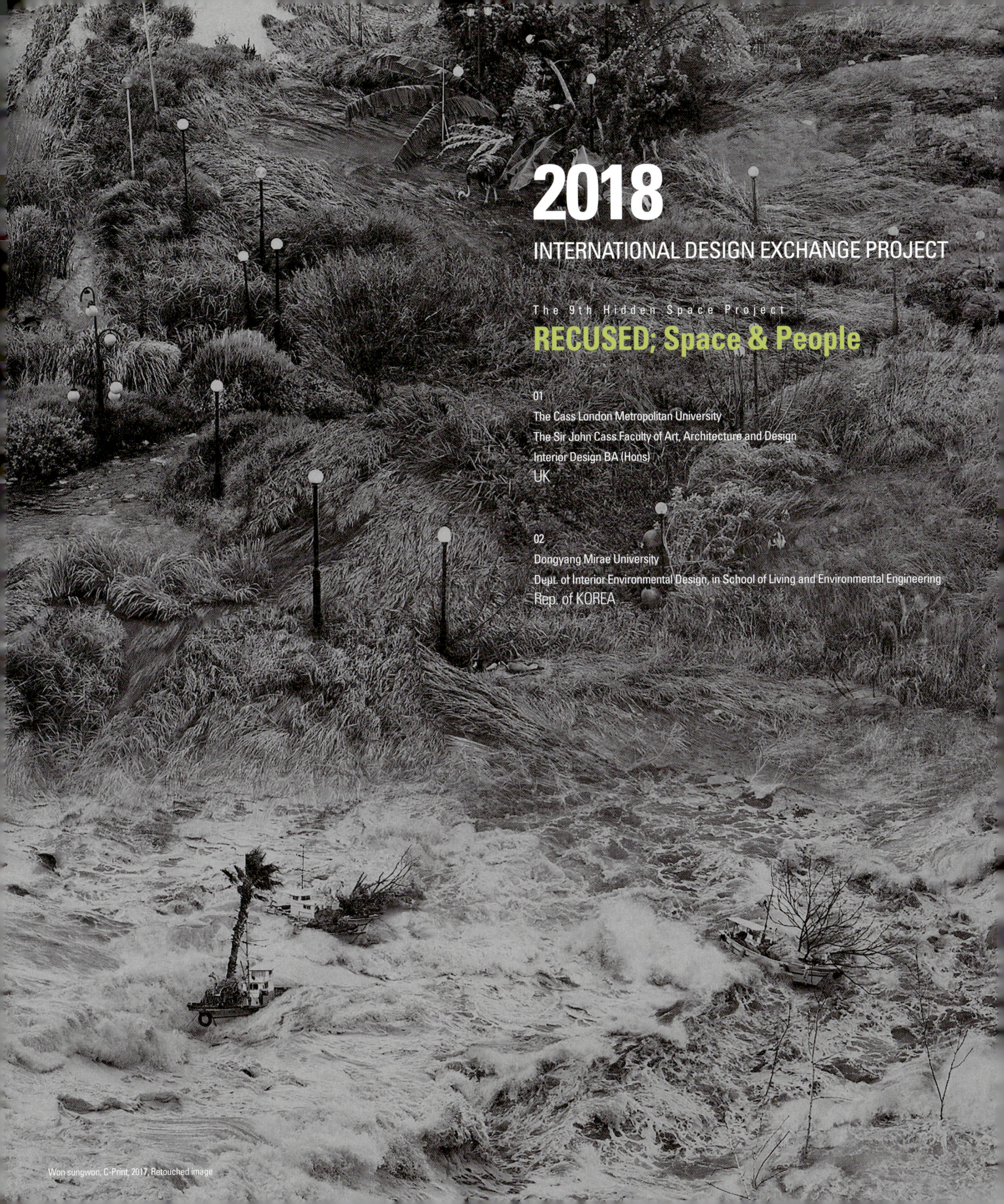

Won sungwon, C-Print, 2017, Retouched image

HOUSE OF PLENTY

_ 16 Projects

contents

01 The Cass London Metropolitan University _UK

02 Dongyang Mirae University _ Rep. of KOREA

INTRODUCTION

PROJECTS

1

The Cass London Metropolitan University
The Sir John Cass Faculty of Art, Architecture and Design
Interior Design BA (Hons)

The Sir John Cass Faculty of Art, Architecture and Design
41-47 Commercial Road
Lonodon E1 1LA
Switchboard 020 7423 0000
www.thecass.com

Kaye Newman MA FRSA

CASS: Head of Interiors
Course Leader: Interior Design BA(Hons)
Acting Course Leader: Interior Design & Decoration BA(Hons)
CASS Faculty of Art, Architecture & Design
London Metropolitan University

Studio 2: London Lost and Found

Cass Head of Interiors: the challenges of curating a brief for the studio that enables thinking of current issues and approaches and allows the student to develop their own practice.

It's never been more challenging to curate a brief through which Interior students strengthen their skills while developing a process and practice. My studio has homed in on Lower Marsh
Lower Marsh which lies just South of Waterloo and dates back to Roman times as it sat on a raised through road cutting through the surrounding Marsh lands adjoining the Cut where is became one of London's longest market streets in medieval times.

The project identifies a cross section of themes and current issues surrounding identity and the make-up of a local High Street. It looks at values of authenticity, character, localism and how this conservation area sits within London itself. The project looks at the significance of living over the shop, and how the space above is as important as the row of shops beneath, where many dysfunctional non connected environments are often situated. The rising needs of suitable accommodation for both a growing older population and in expensive small spaces for the single inhabitant appeared to be paramount.

Lower Marsh sits in the shadow of Waterloo Station and whilst it has suffered the pervading pollution, it has remained quite secret and has avoided over development. However this is now about to change as Waterloo undergoes a revamp restoring the 4 Euro star lines for the South of England where the tunnels and vaults underneath are being developed into retail spaces and eateries to meet the needs of the travellers and other new surrounding developments. This will bring a new customer into Lower Marsh making it a more attractive business opportunity for high street retailers and therefore will provoke the argument of regeneration versus gentrification. Lower Marsh is also home to one of London's oldest street markets and therefore has a very visible vibrancy; it's quirky and has qualities that have been lost in many of high streets.

Last year I was very fortunate to meet with a member of the UN disasters committee and have pondered over our very meaningful conversation ever since. Climate change is now presenting itself within the UK through devastating floods. The Environments agency, via the website ' What's in your backyard' present high risk maps showing areas deluged and overwhelmed by water. The south of London is in great danger of being flooded. Apart from the obvious destruction to property research shows that the lasting emotional damage is significant where behaviour connected to a civilised society can crumble allowing anarchic instances of violence and cruelty to breakout. Therefore the need to be prepared and to build in resilience for such a situation can't be ignored and put off for another day. Lower Marsh gets its name from the time pre 1768 where it was in deed a marsh flanking the Thames and it's this proximity to the river and the tidal forces will add to the overall inevitable destruction.

In order to house all of these issues within one project, I prepared a narrative that asked the students to design a Utopia which suggested that London had suffered one of the worst floods in history; consequently the whole of Lower Marsh was evacuated. Following many weeks the flood subsided to reveal the true picture of damage and devastation. Through the seemingly hopeless challenge to repair and restore a wave of new inspiration emerged. A group of fresh thinking designers offer a new proposal that not only redresses the street and its properties but presents an innovative way of living, rejuvenating livelihoods and the neighbourhood as well as crafting a new direction. Is this a new republic, a new society or a fairy-tale?

It has been such a rewarding year; this scenario has offered the students a raft of mini briefs. The Utopia asks them to set up a programme where the inhabitants live as a collective where there is an inherent support system requiring a sense of neighbourliness. Creating shared spaces and thresholds that encourages friendly behaviour perhaps developing a sense of the extended family. It may not be everyone's cup of tea but how do we start to give the elderly better ideas for integrated living. Similarly student accommodation is proving to be a vast expense in London and in many cases a soulless experience. The whole idea of cohabiting has been explored and how there might be models of living that can be spatially quantified. We have explored ideas that surround moving the retail space up to the upper floors, increasing the business footprint and blurring the boundaries of public and private space, perhaps a family kitchen could also double up as a coffee spot for a bookshop. One of the students has looked at embedding a police house within the inhabited spaces, an old fashioned idea but perhaps custodianship might give the community improved confidence. The studio has developed ideas of flexible furniture that can be used in a multifunctional way for both the Interior and exterior, in a sense of preparedness for the next flood where livelihoods could endure this short-term situation.

The studio are still developing their ideas but I feel that the job of the Interior designer is so important to make these human connected spaces and experiences effective and meaningful, where I would argue that we should be considered as one of the caring industries.

Kaye Newman
Course Organizer - 코스 기획자

Studio 2: London Lost and Found

현재 이슈와 접근법을 생각할 수 있게 해주고 학생이 자신의 실습을 개발할 수 있게 해주는 스튜디오용 큐레이터의 과제.

인테리어 학생들이 과정과 실습을 개발하는 동안 자신의 기술을 강화하는 간단한 과정을 관리하는 것은 결코 쉽지 않다. 내 스튜디오는 Lower Marsh, 더 낮은 습지는 Waterloo의 바로 남쪽에 있으며, 로마 시대로 거슬러 올라간다. 중세 시대에 런던의 가장 긴 시장 거리 중 하나가 된 Cut과 인접한 주변의 습지대를 가로 지르는 도로를 통해 자랐다

이 프로젝트는 주제의 횡단면과 신원을 둘러싼 현안 및 지역 하이 스트리트의 구성을 식별한다. 진정성, 인물, 지역주의의 가치와 이 보전 지역이 런던 자체에 어떻게 자리 잡고 있는지 살펴본다. 프로젝트는 상점을 통해 사는 의미와 위의 공간이 여러 가지 기능이 없는 비연결 환경이 놓여있는 상점 아래의 상점만큼 중요한지 살펴본다. 증가하는 노인 인구와 단일 거주자를 위한 값 비싼 작은 공간 모두에 적합한 적응의 필요성이 가장 중요하게 보였다.

Lower Marsh는 Waterloo Station의 그늘에 자리 잡고 있으며 퍼져있는 오염으로 고통을 겪어 왔지만 아직 비밀리에 남아 개발을 피했다. 그러나 워털루가 터널과 지하실이 여행자 및 기타 새로운 주변 개발요구를 충족시키기 위해 소매 공간과 식당으로 개발되고 있는 영국 남부의 4개 유로 스타 라인을 복원하는 개조공사가 진행됨에 따라 이제 변경 될 예정이다. 이것은 신규 고객을 Lower Marsh에 데려다 줄 것이며, 이로써 번화가의 소매업체에게는 더 매력적인 비즈니스 기회가 될 것이므로 재건 vs 진화에 대한 논쟁이 촉발 될 것이다. Lower Marsh는 또한 런던에서 가장 오래된 거리시장의 본거지이기 때문에 눈에 잘 띄는 활기가 있다. 기발하고 많고 높은 거리에서 잃어버린 자질을 가지고 있다.

작년에 유엔 재난위원회 위원과 만난 것은 매우 행운이었고, 그 이후로 우리의 매우 의미 있는 대화를 생각해 보았다. 기후 변화는 이제 막대한 홍수로 영국 내에서 나타나고 있다. 환경 기관, 웹 사이트를 통해 '당신의 뒷마당에는 무엇이 있습니까?' 지역이 파고 들고 물에 압도 당했다. 런던의 남쪽은 홍수의 위험이 크다. 재산 연구에 대한 명백한 파괴와 별개로, 지속적인 피해는 문명사회와 관련된 행동이 무너질 수 있으며 무정부 상태의 폭력과 잔인함을 조장 할 수 있기에 중요하다. 그러므로 그러한 상황을 준비하고 필요로 하는 탄력성을 무시하고 다음 날로 연기 할 수는 없다. Lower Marsh는 템즈 강을 건너는 습지에 있었던 1768년 이전부터 그 이름을 얻었고 강에 인접해 있어 조력에 의해 피할 수 없는 파괴를 불러올 것이다.

이번 프로젝트에서 이러한 모든 문제를 해결하기 위해 런던에서 역사상 최악의 홍수 중 하나를 겪었다고 제안한 후 유토피아를 디자인하도록 학생들에게 요구하는 서술을 준비했다. 결과적으로 Lower Marsh 전체가 대피했다. 수 주일 만에 홍수에 가라 앉아 손상과 황폐화의 진실된 모습을 드러냈다. 새 영감의 물결을 고치고 복원하고, 겉으로는 희망없는 도전을 통해 나타났다. 새로운 사고방식의 디자이너 그룹은 거리와 그 속성을 시정할 뿐만 아니라 혁신적인 삶의 방식, 생계를 되찾은 생계와 이웃 및 새로운 방향을 제시하는 새로운 대안을 제시한다. 이것이 새로운 공화국, 새로운 사회, 또는 동화일까?

그것은 그토록 보람있는 해였다. 이 시나리오는 학생들에게 미니 보고서의 뗏목을 제공했다. 유토피아는 주민들이 이웃에 대한 감각을 필요로 하는 고유한 지원 시스템이 있는 공동체로서 살고 있는 프로그램을 설립하라고 요구한다. 친숙한 행동을 장려하는 공유공간 및 임계값을 만들어 확장된 가족에 대한 감각을 길러준다. 모두의 차이가 아닐 수도 있지만, 어떻게 보면 노인에게 통합된 생활에 대한 더 나은 아이디어를 제공하기 시작할 것이다. 마찬가지로 학생기숙사는 런던에서는 광대한 비용으로, 많은 경우 영혼없는 경험을 제공한다. 동거에 대한 전체적인 생각이 탐구되었으며, 어떻게 공간적으로 정량화 될 수 있는 삶의 모델이 있을지 모른다. 우리는 소매 공간을 상층으로 이동시키고 사업영역을 확대하며, 공공 및 사설 공간의 경계를 모호하게 하는 아이디어를 모색했다. 아마도 가족용 주방도 서점의 커피 장소로 두 배 더 커질 수 있다. 학생들 중 한 명은 거주지에 경찰관을 임명하는 것을 보았다. 구식 아이디어이지만 아마 관리인이 커뮤니티에 개선 된 자신감을 줄 수 있다.

스튜디오는 생활하면서 단기적인 상황을 견딜 수 있는 다음 홍수에 대한 대비의 의미에서, 내부 및 외부 모두 다기능 방식으로 사용할 수 있는 유연한 가구에 대한 아이디어를 개발했다. 여전히 우리는 아이디어를 개발하고 있지만 실내디자이너의 직업이 이러한 인간의 연결 공간과 경험을 효과적이고 의미 있게 만드는 데 중요하다고 생각하며, 우리가 돌보는 산업으로 간주되어야 한다고 생각한다.

Ellie Boggans
Interior Architecture and Design
Studio 2 NEXUS / Year 3

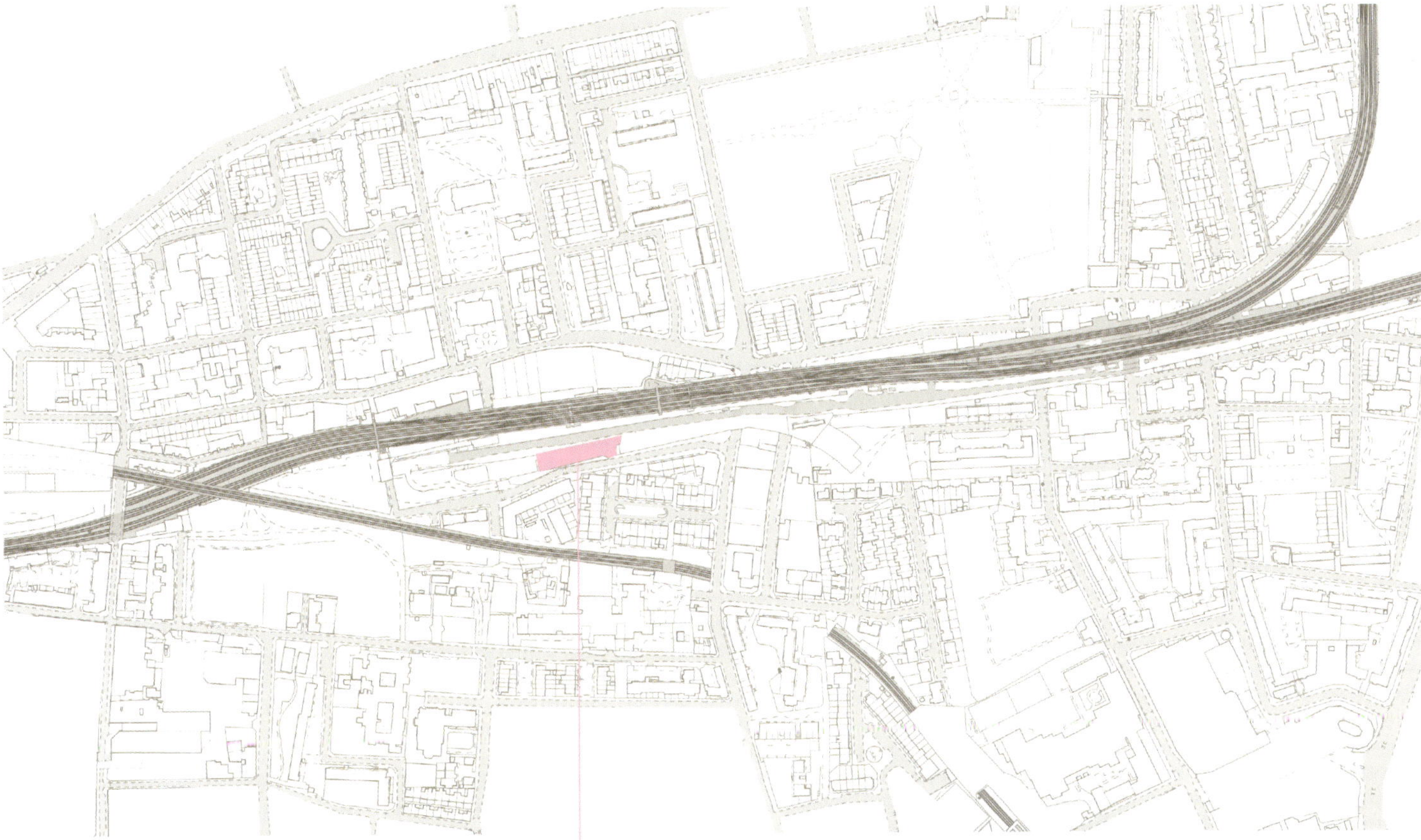

Depot entrance

Depot interior

High-line

Surrounding building

The projects brief is to design a temporary home for a displaced person or persons.

The projects site is Pedley Street depot which is in the Tower Hamlets borough in East London. The depot was a goods yard that was used to store good that were brought in on the railway and is currently still used for storing.

The building is owned by Network rail as there is a railway line that runs on top of the building, leading towards Bethnal Green and is situated in a conservation area.

The access to the can be made directly by walking, cycling or car but only at one end of Pedley Street as it is a dead end. Knowing this means that there wont be any continuous traffic outside the site.

Opposite the site are two building which are esidential housing and the rest of the building on Pedley street at small businesses.

프로젝트의 기본개요는 난민(또는 난민들)을 위한 임시 주택을 설계하는 것이다.

프로젝트 장소는 런던 동부의 타워 햄렛 자치구에 있는 페들리 거리 창고이다. 이 창고는 철도로 들어온 물건들을 보관하는 데 사용되었던 물품 창고였으며 여전히 창고의 역할을 하고 있다.
네트워크 레일이 소유주인 이 건물의 꼭대기에는 베들린 그린으로 향하는 철도가 있으며 보호구역에 위치해 있다.

이 빌딩은 막다른 골목에 있기 때문에 빌딩으로의 접근은 걷거나 자전거를 타거나 차를 타고 갈 수 있지만 페들리 거리의 한쪽 끝에서만 가능하다. 이것은 이 지역 밖으로는 그 어떠한 지속적인 교통이 있을 수 없음을 의미한다.

이 건물의 반대에는 두 개의 빌딩이 있는데 하나는 주택을 제공하고 있으면 나머지 다른 하나의 빌딩은 페들리 거리에서의 작은 상점들이 입주해있다.

Existing

Existing plan shows the space within the depot that we have in the building to work with.

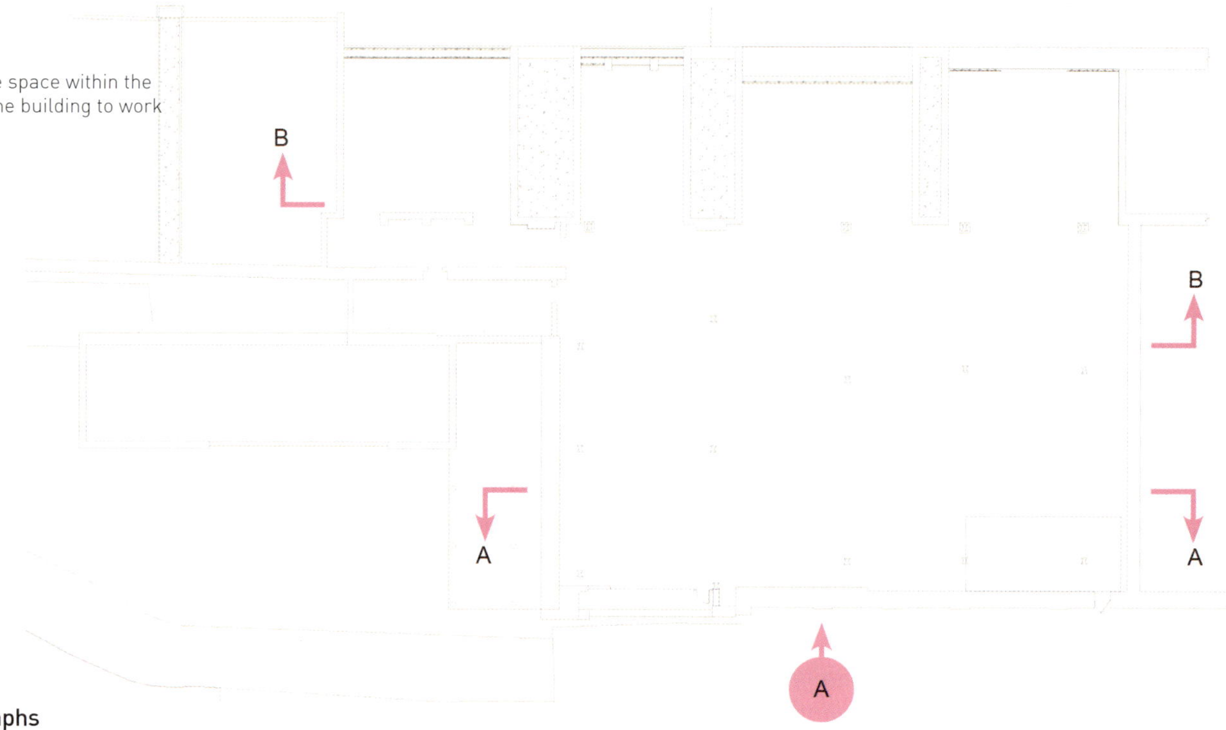

Interior photographs

Arches and some of the interior have Aluminium cladding attached.

The materials in the depot are very industrial, the materials over time have started to corrode and decay to the very little maintenance of the building.

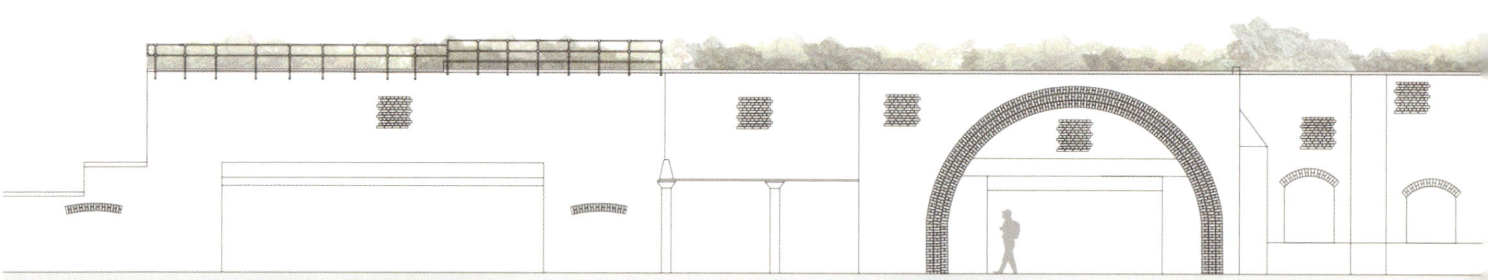

Elevation A - Exterior photographs

Security fencing surrounding depot.

Depot arch entrance.

Existing interior sections.

Section AA

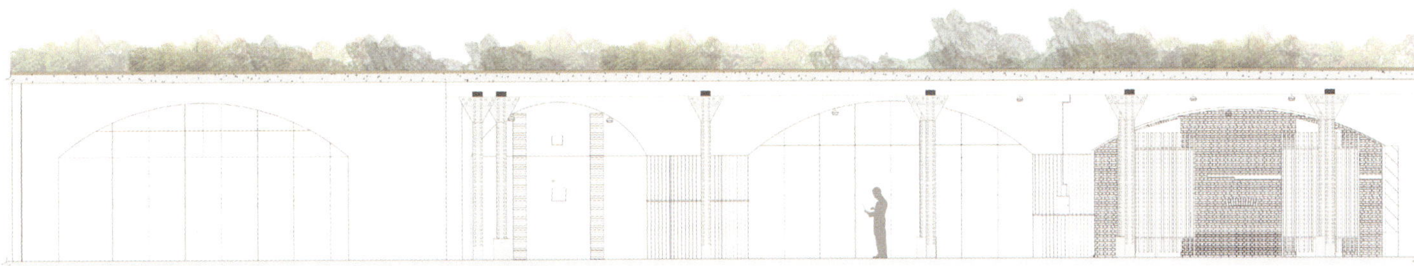

Section BB

The structure holds up the high-line which is situated above, this needs to be strong enough for vehicles and traffic to pass along. Many areas show character of the past but also show a great deal of decay.

Jack arch brick ceiling structure with cast iron beams.

Cast iron columns with rivets.

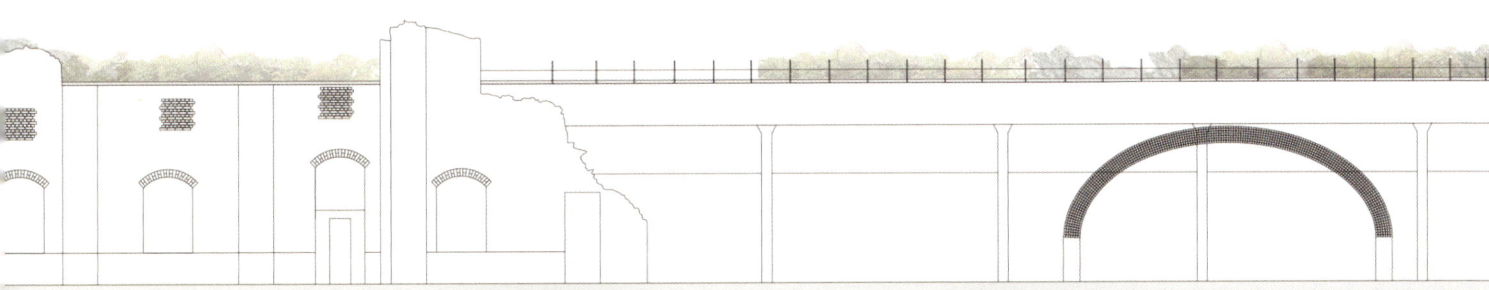

Three brick array.

Outdoor car park area.

Research

The aim is to create temporary community like space for victims of flooding.
The brief asks to design a space that gives people the opportunity to socialise with each-other. In the space there will be a hair dressers, well-bring area, a photo centre, a cafe and other social areas that will allow people to come together and feel comfort during their time of displacement.
There will also be privacy points within the space if people need to have a moment alone or to make important phone calls.
Different types of mapping will help lead to the design of the space.

이 프로젝트는 사람들에게 서로 어울릴 수 있는 기회를 주는 공간을 디자인해 달라고 요청하고 있다. 이 공간에는 미용실과 여전히 과거의 모습을 잘 지니고 있는 지역으로서 사진센터, 카페, 그리고 사람들이 지낼 곳이 없는 기간 동안에 편안함을 느끼고 함께 즐기는 것을 가능하게 해 주는 다른 사회공동체적인 구역이 있다.
사람들이 중요한 전화를 하거나 혼자만의 시간을 필요하다면 이러한 공간 내에 사적인 부분이 또한 있을 것이다. 다른 유형의 매핑은 공간 설계를 이끄는 데 도움을 준다고 볼 수 있다.

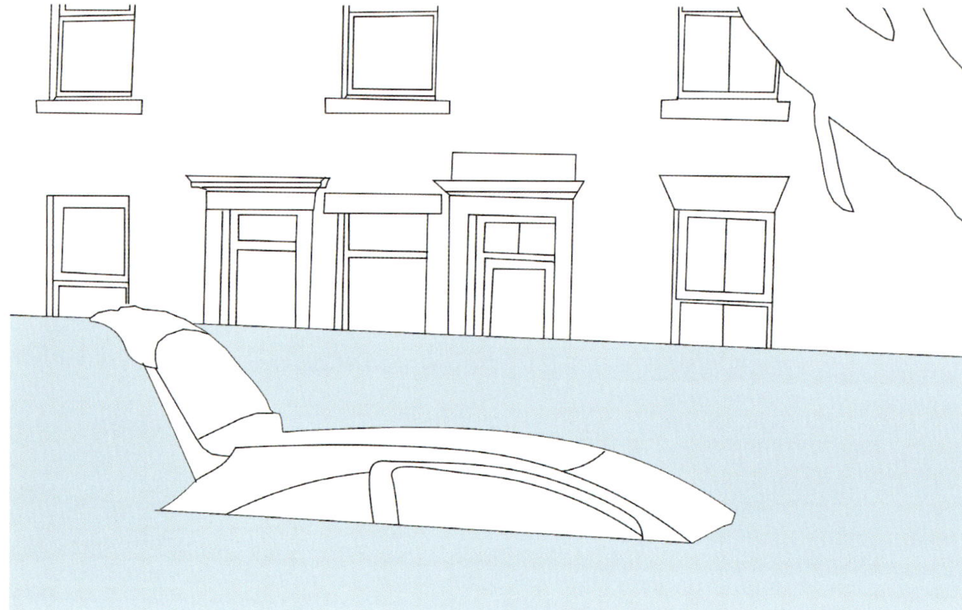

"It was waist-deep. The time I went out in it to get the last few bits I could save, it was up to my waist. The last time I went in the house the fridge-freezer was on its side, the bins were all out, the TV was damaged. Everything was just under water, it was that bad."

"It's affecting my health. I know I'm drinking more than I used to. When you're not comfortable and you're not in your own home for a second Christmas it's not much fun."

"The traffic was making the flooding worse in your house. We tried to stop it but you just gave up because it kept coming in."

"허리까지 차 있었다. 나를 구할 수 있는 마지막 탈출하는 그 순간, 이미 허리까지 물은 올라와 있었다. 그 후 지난 번 내가 집에 갔을 때 냉장고는 한 쪽 구석에 있었고 쓰레기통은 모두 나뒹굴고 TV는 고장이었다.모든 것이 물 속에 있었으면 그것은 아주 끔찍한 것이다."

"이것은 나의 건강에도 영향을 주고 있다. 과거의 내 자신보다 더 많은 술을 마시고 있다. 편안하지 않고 당신 자신의 집도 아닌 곳에서의 두 번째 크리스마스는 즐거운 일이 아니다."

"차량들은 이러한 홍수를 더욱 끔찍하게 만들었다. 그것을 멈추게 하려고 노력했지만 계속적으로 떠밀려 내려오는 그것들에 우리는 단지 포기를 할 수 밖에 없었다."

Based on research from articles and personal interviews from victims of flooding it was clear that the victims were unable to get their personal belongings during the flood so it is important for them to have a place where they can make arrangements.

Images are from the devastation caused by storm Desmond in late December 2015. The storm affected UK and Ireland and caused millions of pounds worth of damage.

홍수 피해자들의 기사와 개인 인터뷰 조사에 따르면, 피해자들이 홍수 동안 개인 소지품을 구할 수 없다는 것이 분명했기 때문에 그들이 준비할 수 있는 장소를 갖는 것이 중요하다.
이미지들은 2015년 12월 말에 폭풍 데스몬드 때문에 생긴 혼돈을 보여주는 것이다. 그 폭풍은 영국과 아일랜드에 영향을 미쳤고 수백만 파운드의 피해를 입혔다.

The different size and colour of each circle represents the noise levels through out the house from the same spot.
From doing this experiment we can see that cleaning is a continuous routine throughout the day at all different types. From this information it is important that cleaning is taking into consideration when designing the pod space for the displaced person/ persons.

각 원들의 다른 크기와 색깔은 같은 지점으로부터 집안내에서의 소음 수준을 나타낸다.
이러한 실험을 통해서 우리는 청소가 각양각색의 지속적인 일상임을 알 수 있다. 이러한 정보를 통해서 갈 곳을 잃은 사람들을 위한 pod공간을 설계할 때 청소를 고려하는 것이 중요한 요소이다.

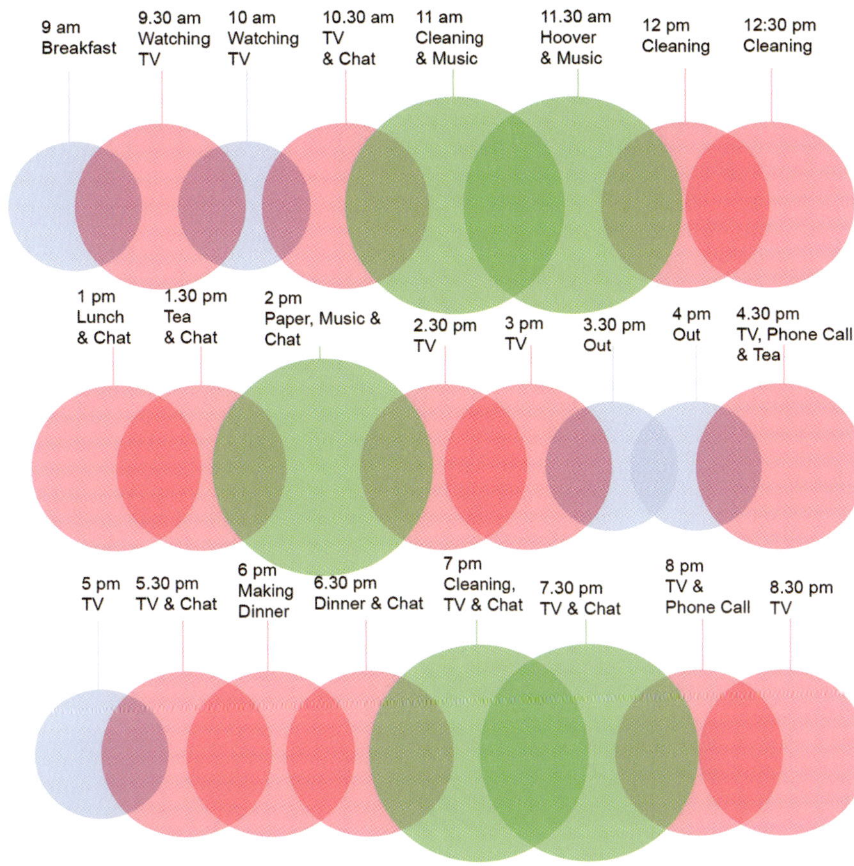

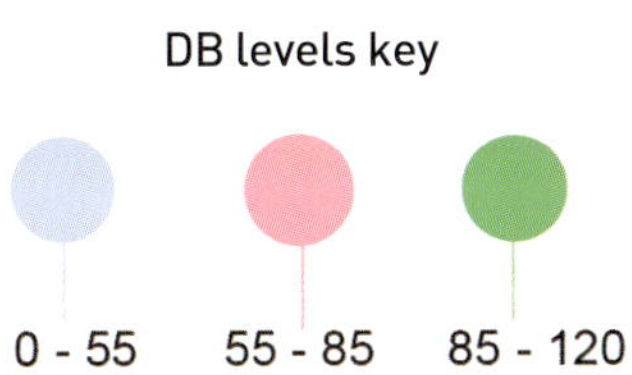

The many uses of The kitchen table.
The table belongs in a home where a older woman lives with her partner and granddaughter. The table is a social hub in the home when people come over to visit.

주방테이블의 많은 용례들.
나이가 지긋이 든 여성이 그녀의 배우자와 손녀와 함께 사는 집에 테이블이 있다. 이 테이블은 사람들이 방문할 집안에서의 사회적인 중간 연결고리이다.

- Socialising
- Eating
- Uni work
- Chatting
- Children colouring
- Art work
- Reading

Routine Mapping

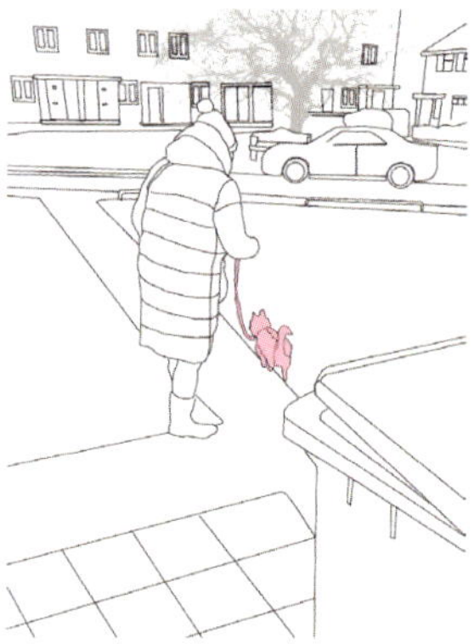

Dog walking

Cleaning/Hoovering - 85 - 120 DB

Eating Lunch - 55 - 85 DB

Cooking Dinner - 55 - 85 DB

Dinner - 55 - 85 DB

Watching TV - 55 - 85 DB

From mapping the routine of the displaced person it shows that dog walking is an important factor, it would be good to provide an area that people could take walks whether it be with their pets or not.

갈 곳을 잃은 사람들의 일상생활 속 매핑으로부터 개 산책이 중요한 요소임을 알 수 있고 애완동물과 함께 있든 아니든 사람들이 애완동물과 걸을 수 있는 장소를 제공하는 것이 좋은 것이다.

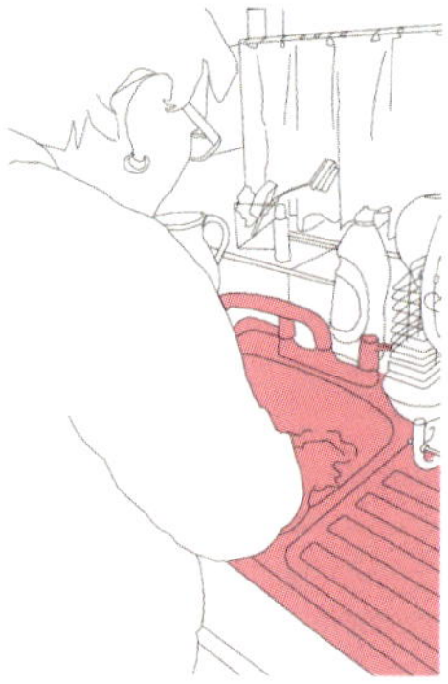

Washing up - 85 - 120 DB

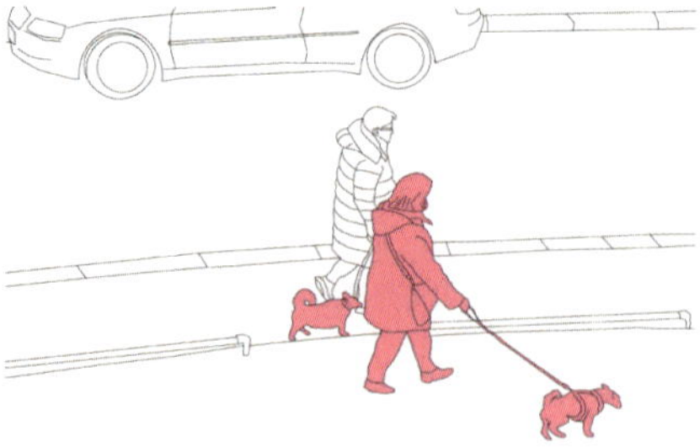

Dog walking

Hero Page

A representation of how the table can bring people together to create a community.

Proposed Elevation & Section

The greenery and trees at the front of the depot make it more welcoming to both the public and pod residents.

The high-line can be a place where people can socialise, walk their dogs, having barbecues in the summer and to also do some gardening.

Pods on the high-line is a space for residence only and wont be able to be accessed by the public.

Well Being Zone: The private well-being zone allows people to have a quite space away from the crowded depot to relax and clear their mind. The space includes well being pods which allows people to climb inside and isolate themselves from the outside to give themselves a moment of privacy.

Coffee Stop: The coffee shop in the depot is a social point in the site as hot drinks supposedly give a feel of comfort. As in my house a cup of tea is always on offer in times of distress.

House of plenty: Main entrance to the depot, will be lighted up to show exist-ence and let in be known that its there.

Adding windows to the building with make the space brighter and more open as it is currently very dark as there are no openings other then the main entrance. This is also in the suns direction so a lot of sun will enter the building.

Adding plants onto the high line helps to purify the air, this will help as the railway lines are right next to it.

Hairdressers: The hairdresser will be a part of the depot that will allow both residents of the pod and the public to come. From research I chose a hairdressers because it was described as a 'get away' from thinking about stressful situation and a place to socialise and relax.

Photo station: The photo station is a hub in the depot that allows pod residents to print off photographs which may have been lost during the disasters.

Herb table allows users of the space to grow herbs and vegetables. This will be a social point within the space.

Hanging seating

A study area has been designed into the space, areas in the space are both private and social areas.

Phone-booths create a space for private/important calls to be made. The booths will be lined with hush foam baffles to ensure that conversation cannot be heard.

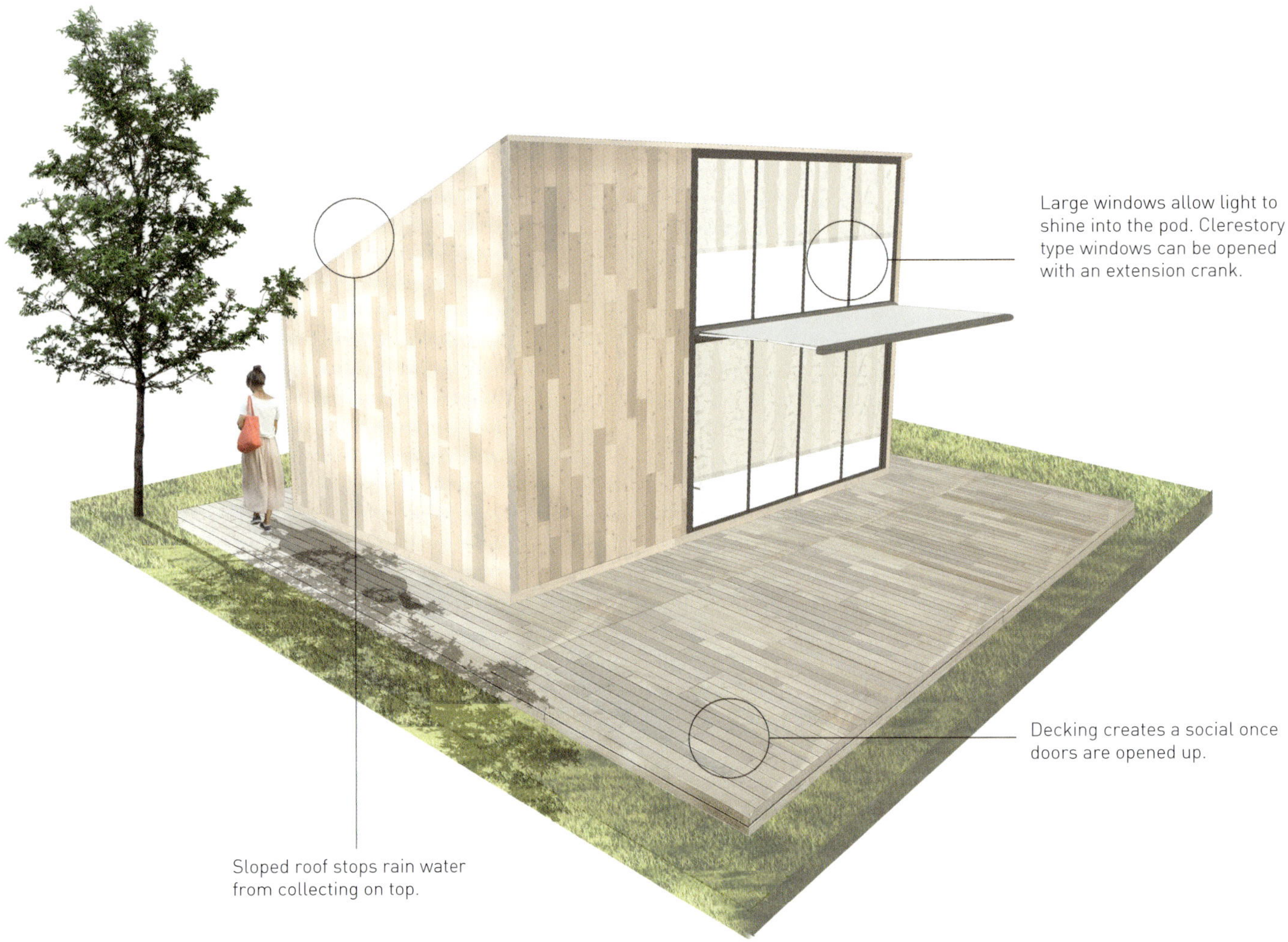

The images and diagrams show the proposed plan for the displaced person or persons pod design. The Pod design reflects on the family mapping and research that was gathered. A social area in the pod (the decking) allows the pod residents to socialise with other residents.
The bi-fold doors and the transforming furniture was a way of creating a mix space, social and private. The plans show the spacial analysis of the pod.

이미지 및 다이어그램은 갈 곳을 잃은 사람들의 pod공간설계에 대해 제안된 계획을 보여주고 있다. . pod 공간은 수집된 조사와 가족 매핑을 반영하고 있다. pod(decking)의 사회적 영역에서는 pod거주자들이 다른 거주자들과 어울릴 수 있다. 양면 문과 변형할 수 있는 가구는 사회적이고 사적인 혼합된 공간을 만드는 한 가지 방법이었다. 제시된 계획들은 pod의 특별한 분석을 보여주고 있다.

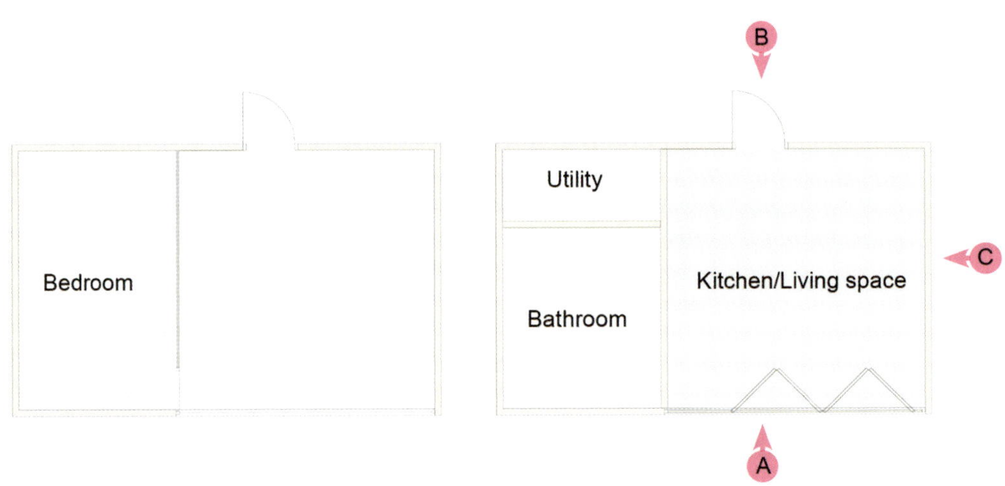

Proposed Pod Design

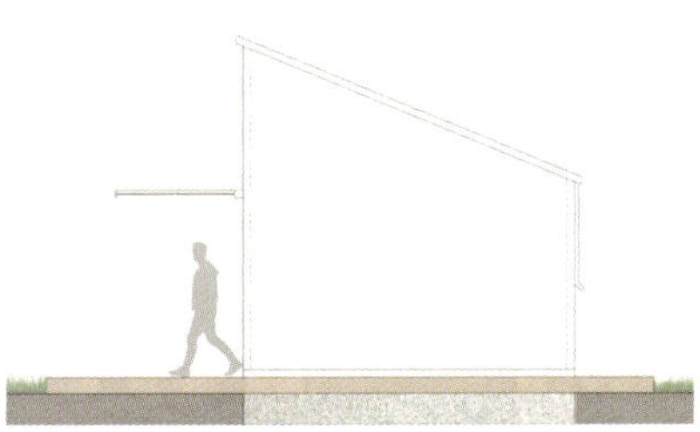

Elevation A

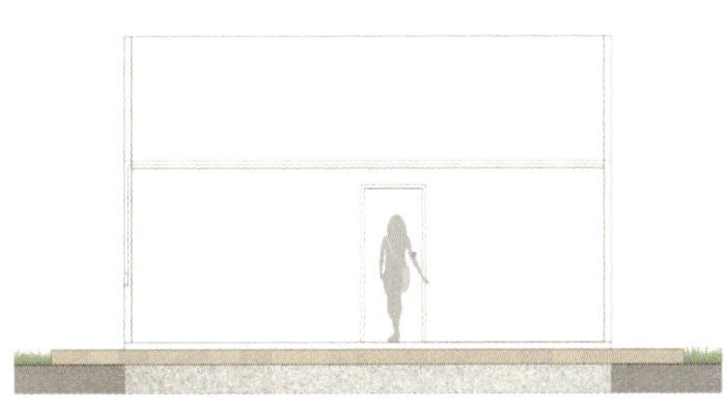

Elevation B

Elevation C

Following page shows the transformative piece of furniture you can see in the visual above (the table on decking).
다음 페이지는 시각적으로 볼 수 있는 변형 가능한 가구 모습을 보여주고 있다 (decking의 테이블).

Transformative furniture

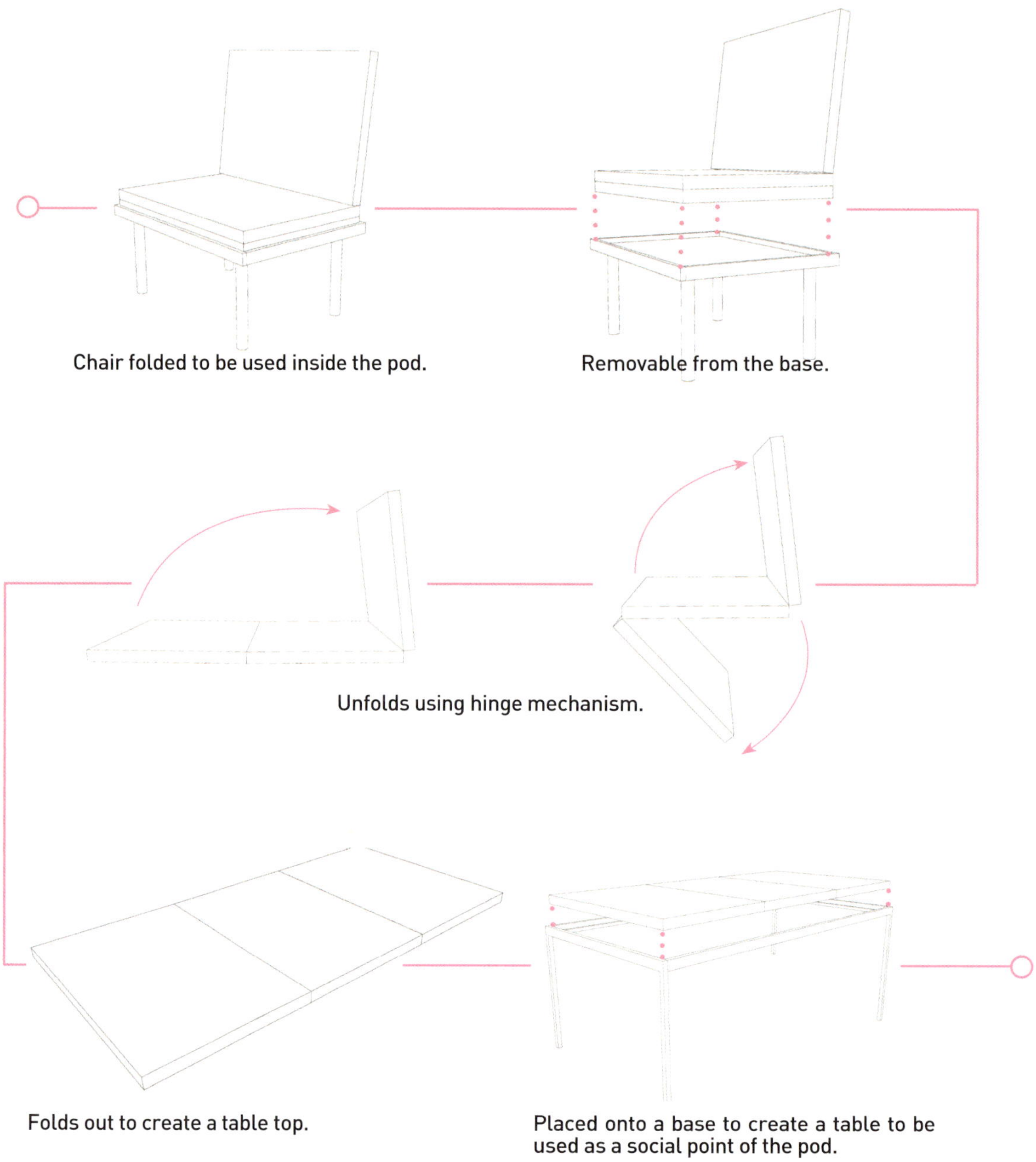

The chair will be a piece of furniture for the inside of the pod but can be transformed into a table which can be used as a socialising point for the pod. This will allow the neighbouring displaced to interact with each-other and come together as a community.

Waterproof tropical timber with a powder coated cast iron base.

Proposed Plan

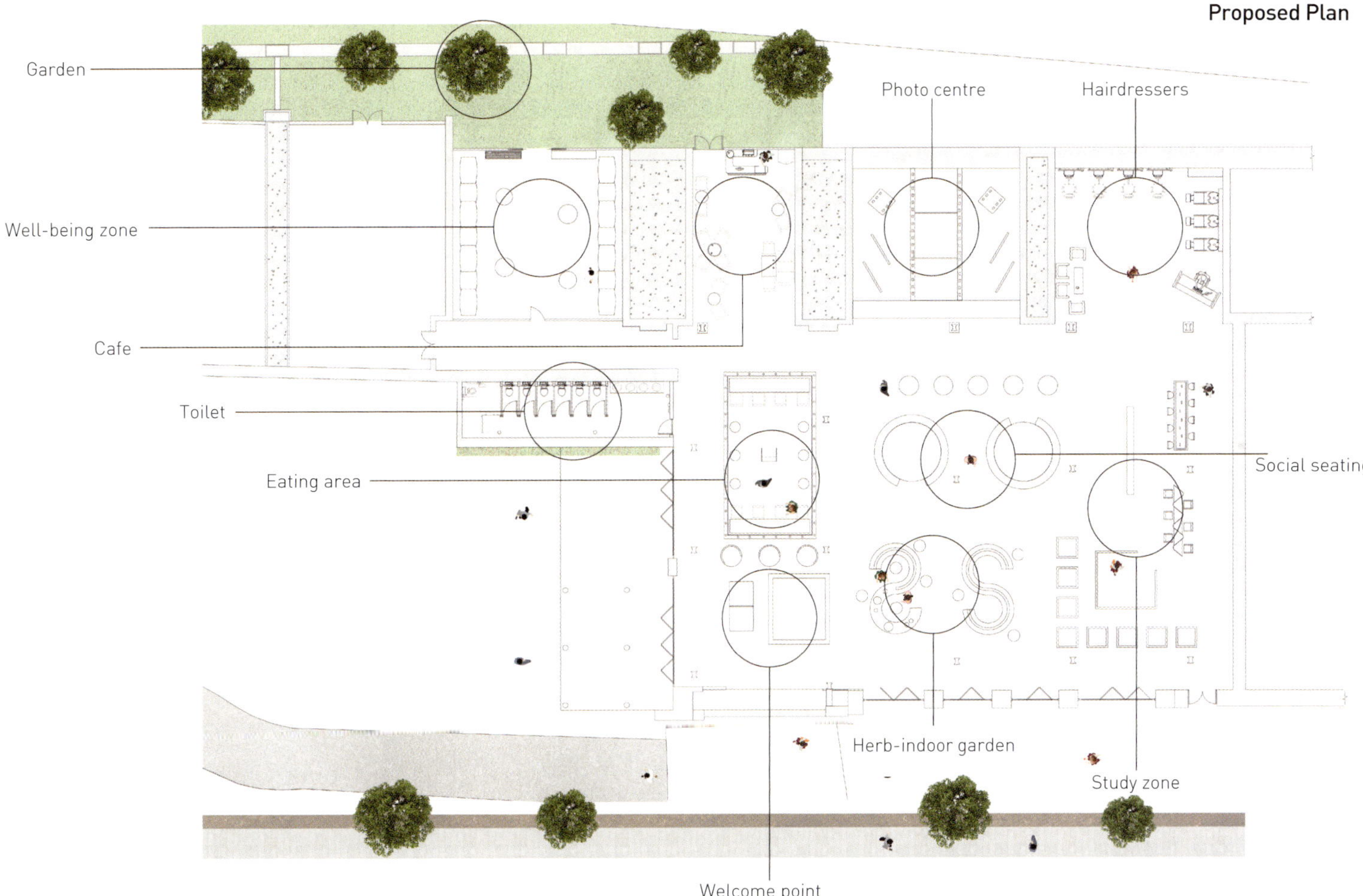

Plan show the proposed plan of the depot. The interior of the depot is where the business will be and can be used by the public and pod residents. The depot provides a space for community to come together.

제시된 계획들은 제안된 창고 계획을 보여준다. 창고 안에서는 비즈니스가 있을 수 있고 대중과 pod거주자들이 사용할 수 있다. 창고는 공동체가 함께 모일 수 있는 공간을 제공한다

Plan showing high-line and surrounding buildings. The high-line will be where the pods are placed and is a private area for pod residents.
The high-line also provides the residents with an area where they can walk their dogs.

제시된 계획들은 높은 나무지대와 주변 건물들을 보여준다. 높은 나무지대에는 pods가 위치하고 pod거주자들을 위한 개인적인 공간이 될 것이다.
또한 이러한 높은 나무지내에는 거주자들이 그들의 애완동물과 걸을 수 있는 공간이 제공 될 것이다.

Laura Veronica Grieco
Interior Architecture and Design
Studio 2 NEXUS / Year 3

NEXUS
PROJECT

"Nexus" project focuses on thinking of homes for the displace individual, how this can foster community and sense of home.
By researching and studying the context area of the site and the depot located in Pedley street, brief requires to look at new occupation, the re-invention of the interiors to accommodate the functional, emotional and responsible needs of the inhabitant.

The aim of "Nexus" is to demonstrate a community's response to crisis and immediacy.

Following research project focuses on developing a home for displaced women and child escaping from abusive homes. Emphasising on the wellness through the contact with nature, offering spaces for reflection, peace and growth.

"Nexus" 프로젝트는 대체 개인을 위한 집을 생각하는 것, 이것이 어떻게 지역사회와 집처럼 느낄 수 있는 감정을 키울 수 있는가에 초점을 맞추고 있다.
페들리 가에 위치한 부지 및 창고의 지역을 조사하고 연구함으로써, 거주자의 기능적이고 정서적이며 책임 있는 요구를 수용하기 위해 내부가 새롭게 생겨나는 것을 살펴볼 필요가 있다.

Nexus의 목적은 위기와 신속성에 대한 지역사회의 반응을 보여주는 것이다.

아래의 프로젝트는 갈 곳을 잃은 여성과 학대하는 가정으로부터 탈출하는 아이들을 위한 집을 개발하는 것에 초점을 맞춘다. 특히 자연과의 접촉을 통한 웰빙 강조, 반영, 평화, 성장을 위한 공간 제공을 집중적인 조명이 이루어진다.

Site location

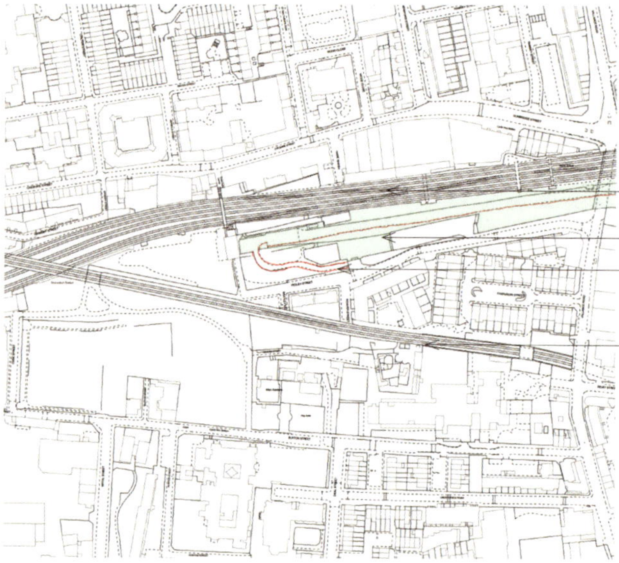

1.

2.

3.

1.Interior of arches running along Bishop Goodyard
2.Bishopsgate Good Station September 1995
3.Bishopsgate Good Station January 2018

Bishopgate interior arches carries the same construction as Pedley street depot, including jack arches, cast iron beams, train track embedded on the floor. This area of Bishopsgate as our depot suggest previous basement floor as well as interior arches.

비숍게이트 내부 아치는 잭 아치, 주철 빔, 열차 트랙이 바닥에 박혀 있는 페들리 거리 창고와 같은 구조물을 가지고 있다. 우리가 제안하는 지하층과 내부 아치의 모양은 이것과 같은 모양이다

About the site and its history

Pedley street depot is historically part of the Bishopgate goods yard

Freight trains used to run in through the interior of the series of depots were used for the loading or unloading of full wagon-loads of fruit, potatoes , vegetables and fish.

페들리 거리 창고는 역사적으로 비숍게이트 물품 보관소의 일부입니다.

일련의 창고 내부를 통과하는 화물열차는 과일, 감자, 야채, 생선의 모든 양을 열차를 싣거나 내리는 데 사용되었다.

National Network Railway new ginger line cuts Pedley Street area into a triangle, isolating it from the other areas, therefore more chance of anti social behaviour. The same happens to the area between Nomadic gardens and Allan gardens.

내셔널 네트워크 철도회사의 새로운 생강선은 페들리 스트리트 지역을 삼각지역으로 잘라내서 다른 지역과 분리시켜, 따라서 반사회적 행동의 가능성을 높인다. 유목민정원과 앨런정원 사이에도 같은 일이 일어난다.

Domestic abuse in Tower Hamlets Borough

What is domestic abuse?
An incident of threatening behaviour violence or abuse (physiological, physical, sexual, financial or emotional), between adults who are, or have been intimate partners or family members.
Domestic violence is a significant problem in Tower Hamlets, with the police dealing with an average of 11 incidents every day
Venues and associations across Tower Hamlets keep location discreet or hidden to provide more security and protection. Call centres are used to advice those suffering domestic violence and guide them by offering groups sessions in undisclosed locations.

가정폭력이란 무엇인가?
성인이거나 가까운 파트너이거나 가족 구성원 사이에서의 위협적인 행동 폭력 또는 학대(생리학적, 신체적, 성적, 재정적 또는 감정적)의 사건.
타워 햄릿에서는 가정 폭력이 중요한 문제로 매일 평균 11건의 사건을 경찰들이 이러한 일을 다루고 있고 이 지역 전역에 걸친 여러 단체들은 더 많은 보안과 보호를 제공하기 위해 조심스럽거나 은닉된 장소를 유지하고 있다. 콜 센터는 가정 폭력에 시달리는 사람들에게 조언을 해주고, 공개되지 않은 장소에서 그룹 세션을 제공하여 그들을 안내하는데 사용된다.

How nature supports victims recovery

The "Environmental Design in Support of A Trauma Recovery Program for Female Adolescent Victims..A Courage House for Courage Worldwide, Inc", emphasises the importance of natural lighting as it supports the patient by helping reduce depression and promoting better moods.
Contact with nature has a quick effect on reducing levels of stress, anxiety, anger and also offers patients the environment, the opportunity and space to reflect.

The "Environmental Design in Support of A Trauma Recovery Program for Female Adolescent Victims" 와 "A Courage House for Courage Worldwide, Inc" 에서는 가정 폭력의 환자들에게 우울증을 줄이고 기분을 좀 더 좋게 해줄 수 있는 자연 조명의 중요성을 강조한다.
자연과의 접촉은 스트레스, 불안, 분노의 정도를 줄이는 데 빠른 영향을 미치며 또한 환자에게 이러한 마음을 반영할 환경, 기회, 그리고 공간을 제공한다.

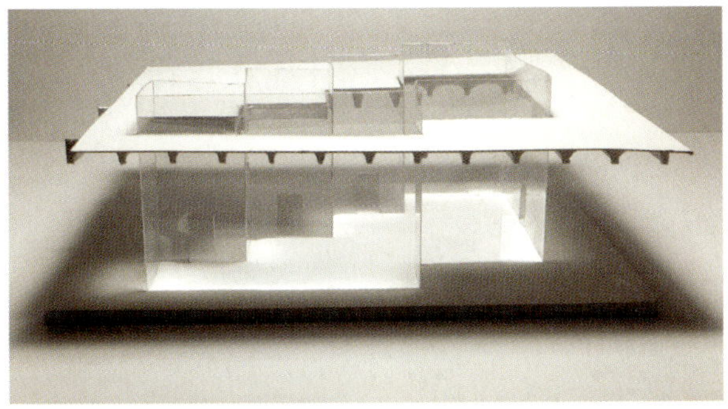

Indoor open roof garden design which works best with the jack arches roof construction, the indoor garden with roof structure model shows the relationship with both as well as showing how will it be seen from the high-line roof.

잭 아치 지붕 구조에서 가장 잘 작동하는 실내 오픈 루프 정원 디자인, 지붕 구조 모델이 있는 실내 정원은 두 가지 모두와의 관계를 보여 주며, 또한 하이 라인 지붕에서 어떻게 보일지 보여준다.

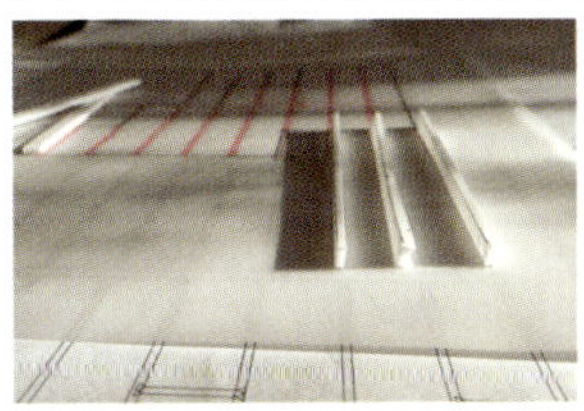

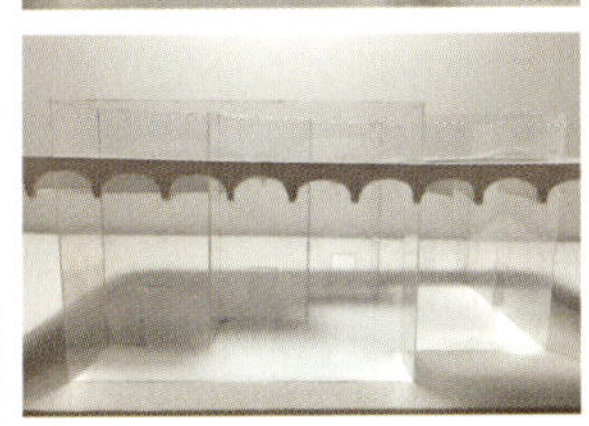

Rubber Plants

Combats:

- Carbon monoxide
- Formaldehyde
- trichloroethylene

They require little maintenance

Being the centre of the indoor garden, offering its beautiful colours to better moods, Cherry blossom also benefits health largely because it contains antioxidant and anti-inflammatory properties.
Studies show that Cherry Blossom extract from Prunus lannesiana revealed its function on advanced glycation-end products (AGE). Age's (a befitting acronym) are toxins that promote aging, inflammation and oxidative stress that lead to other deceases.

더 나은 분위기에 아름다운 색깔을 제공하는 실내정원의 중심인 벚꽃은 또한 항산화제와 항염증 특성을 포함하고 있기때문에 건강에 많은 도움이 된다.
연구는 Prunus lannesia의 벚꽃추출물이 고급 글리제이션 엔드 제품(AGE)에 대한 기능을 나타낸다는 것을 보여준다.
연령(적합한 약어)은 노화와 염증 그리고 다른 사망을 유발하는 산성화 스트레스를 촉진하는 독소이다.

Snake plants

Combats:

- Carbon monoxide
- Xylene (found in petrol and rust preventers)
- Formaldehyde (emitted with vehicle exhausts)
- Trichloroethlene
- Toluene
- Benzene

Best in cool temperatures and indirect sunlight

English Ivy

Combats:

- Benzene
- Carbon monoxide
- Formaldehyde trichloroethylene

They require little maintenance

Peace Lilly

Combats:

- Benzene
- Trichloroethylene
- Formaldehyde toxins
- Boost a room's humidity by up to 5%

These plants are easy to care for

PROGRAM AIMS

Centre aims to support younger women and their children to escape and recover, from traumatic and dangerous home abuse experience. This is based on the high level of cases currently taking place in Tower Hamlets Borough.

센터는 젊은 여성들과 그들의 아이들이 충격적이고 위험한 가정 학대 경험으로부터 탈출하고 회복하도록 지원하는 것을 목표로 한다. 이것은 현재 타워 햄릿 자치구에서 발생하고 있는 높은 수준의 사건들에 바탕을 두고 있다.

Victims will receive information about the centre through police and hospital staff. To continue victims will safely visit the centre's natural cafe (access to cafe by membership only) informing their obsessive partners that they will go for a coffee and bring the child to the cafe's nursery. When in reality whilst child plays in nursery the mother's meet an undercover centre's staff member who discretely discuses with the victim the situation and offers advice.

피해자들은 경찰과 병원 관계자들을 통해 센터에 대한 정보를 받을 것이다. 피해자가 안전하게 이 센터의 자연 카페(회원으로만 카페에 접근할 수 있음)' 을 방문해서 그들의 비정상적인 강박증에 시달리는 파트너의 정보를 주고 이러한 카페에서 차를 마시거나 아이들을 카페의 탁아소에 데리고 올 수 있다. 실제로 아이가 보육원에서 노는 동안 엄마들은 보호소의 직원들과 만나게 되는데, 이들은 그 상황을 피해자와 별개로 토론하고 조언을 해준다.

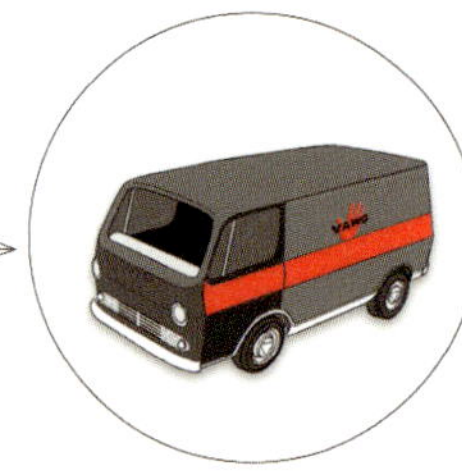

After securely arranging with the centre's staff details of when, where and how, the centre's disguised van will pick up the victim with her child and transport secretly to the centre.

언제, 어디서, 어떻게 하는가에 대한 센터의 직원 세부 사항들을 완전하게 정리한 후, 센터의 위장 밴이 그녀의 아이와 함께 피해자를 데려와서 몰래 센터으로 수송할 것이다.

The van will make its way through Fleet Hill street tunnel, where the centre's secret back entrance is located. Allowing the victims and their child to access the centre safely without anyone knowing where they are or that this centre even exist in Pedley street.

이 밴은 이 센터의 비밀 뒷문이 있는 플리트 힐 거리 터널을 지나가게 될 것이다. 피해자와 아이가 어디에 있는지, 심지어 이 센터가 있는지 조차도 아무도 모르게 안전하게 센터에 접근할 수 있도록 가능하게 해 준다.

The victims will go through regular check ups and physical test to make sure any possible injury is treated correctly.

피해자들은 어떤 부상이라도 제대로 치료하기 위해 정기 검진과 신체 검사를 받게 될 것이다.

Clothes and every day essentials will be provided upon arrival, as many will escape their homes without their belonging to avoid their abuser becoming suspicious.

많은 사람들이 학대자가 의심받는 것을 피하기 위해 그들의 필수 소유물 없이 집에서 탈출할 것이기 때문에, 도착 시 옷과 매일 필수품들이 제공될 것이다.

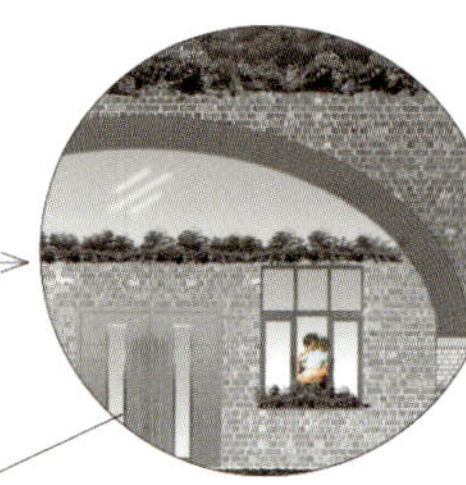

A secret temporary accommodation within the centre will also be provided. The period of stay will depend on the victims' speed progress. As the accommodation will run in stages, first stage up to 2 victims will be sharing a home. Second stage, once mothers have shown progress in recovery from injury, depression, fear and any other negative emotional symptoms they will be relocated to the pod housing. This stage will support them in becoming used to and be able to live at home with their child, as well as a step to becoming independent secure women before moving out of the centre and start a brand new happy life with their child.

센터 내의 비밀스러운 임시 숙소도 제공될 것이다. 체류기간은 피해자의 진전 속도에 달려 있다. 숙박시설이 단계적으로 운영되기 때문에 1단계는 최대 2명의 피해자들이 함께 살게 된다. 두 번째 단계에서는, 어머니들이 부상, 우울증, 공포, 그리고 다른 부정적인 감정적 증상에서 회복하는 치료 차도에 따라 Pod 주택으로 옮겨질 것이다. 이 단계에서 그들이 아이들과 함께 집에서 사는 것이 익숙해지도록 그리고 실제로 같이 살 수 있도록 도움을 줄 것이고 또한 센터를 나가기 전에 자립심이 강한 한 여성이 되는 단계이기도 하고 그들의 아이와 함께 새로운 행복한 삶을 시작하는 단계이기도 하다.

Physiological support and group sessions will be provided within the centre's program. This will support them in gaining back their confidence, express their fears, share with others who are experiencing similar situation and hear inspirational talks by those who have been through this situation and thanks to the centre they liberated themselves and built a happy health life.

생리적 지원과 그룹 세션은 센터의 프로그램 내에서 제공된다. 이것은 그들이 자신감을 되찾고, 두려움을 표현하고, 비슷한 상황을 겪고 있는 다른 사람들과 공유하고, 이 상황을 겪은 사람들에 의해 고무적인 대화를 듣고, 이러한 센터 덕분에 그들이 스스로를 해방시키고 행복한 삶을 만들어 갈 수 있도록 도울 것이다.

Nursery for children will be provided, for those in school age, a home school system carried by voluntary working tutors offering lectures within the each centre's home.

어린이용 보육시설은 학교 연령대 아동에게 제공되며, 각 센터 가정 내에서 강의를 제공하는 자원 봉사 강사들이 운영하는 홈 스쿨 시스템이 제공된다.

As part of the programs the residents will take up lessons in gardening. The connection with nature is proven to lower stress, anxiety, depression symptoms and help better mood. At the same time they would be contributing to the maintenance of the centre, which will be disguised and covered fully by plants, flowers, trees and nature. Besides gardening, once the victims are in the right stage mentally and physically they will use the vegetables and ingredients from the gardens to cook in the centre's cafe. The cafe's earnings will go partially to the centre but also part for the moms. This will help them to be able to learn how to manage finances and save for when they move out of the centre.

프로그램의 일환으로 거주자들은 원예 수업을 받을 것이다. 자연과의 교감은 스트레스, 불안, 우울증을 줄여주고 기분을 좋게 해준다. 동시에 그들은 센터의 유지에 기여하고, 센터는 식물, 꽃, 나무 그리고 자연에 의해 완전히 가려지고 덮일 것이다. 정원 가꾸기 외에도, 일단 희생자들이 정신적, 육체적으로 올바른 단계에 있게 되면 그들은 센터 카페에서 요리하기 위해 정원의 야채와 재료들을 사용할 것이다. 이 카페의 수익은 부분적으로 센터에 전달되겠지만 어머니들 몫이기도 하다. 이것은 그들이 재정을 관리하는 방법을 배울 수 있고 그들이 센터에서 벗어날 때를 위해 저축할 수 있도록 도와줄 것이다.

1 Nature will act as a end way to divide the high line from the centre's area to the rest of high line which runs towards Bethnal Green

2 In the courtyard area tall trees, and bushes will stop commuters from train looking into the residence area. Crawlers plants will cover part of the residence's walls to camouflage the building.

3 Nature will help to combat the pollution created by trains running near by and separate the courtyard area by enclosing it and creating a more private space

4 Cherry Blossom tree is core of the indoor garden allowing viewers to enjoy its beauty from top of high-line into the garden

5 Wild flowers and planting will cover the walls of centre avoiding passers by looking into the centre

6 Small allotment areas will form part of the outside space to grow vegetables and various plantings

7 Wild plants trees and bushes run along the edges of the high line to create a barrier and create a more private space

Floor plan design

1 Main wellness kitchen cafe entrance for controlled public access. Access permitted through membership.

2 Main wellness kitchen cafe area. Here victims secretly meet undercover centre's advisors to discus on how to safely make the transition of leaving home and become residents in the centre.

3 Indoor garden area. This offers a space for reflection, peace and relaxation for the residents victims. It also provides them with nature healing properties.

4 Pod indoor parking area for when they do not wish to have the mobile pod home outside.

5 Communal toilets.

6 Communal laundry area for residents.

7 Centre is camouflaged by plants, flowers and nature. This disguises the real use of the building but also encourages bees, bird and wildlife into the area as well as helping air pollution.

8 Slope with buried tracks allowing stage two mobile Pod house to access high-line where residents live more independently.

9 Green area. Sections are used for growing vegetables and other ingredients to later use in the wellness kitchen cafe.

10 Residents take up gardening lessons as part of their recovery program. Contact with nature diminishes anxiety, depression and promotes better mood. Also this way they contribute to the maintenance of the centre.

11 South facing in and out access from arch residency. Allows residents to make most use of green space which benefits of most sunlight.

12 Wellness kitchen cafe cooking station area. Victims in the stage of being comfortable with other around and became more sociable learn how to cook and work in the kitchen using ingredients from centre's green areas planting.

13 Nursery area. Whilst mothers receive advice from the undercover staff advisor kids stay in nursery learning and playing.

14 Treatment room where victims go through a check ups upon arrival but also assesses their health on regular basis.

15 Group and one to one session area. Victims will attend regularly as part of recovery progress lead by professionals volunteers.

16 Centre's staff members office area and break room area.

17 First stage housing, victims residents share accommodation as part of the beginning of their recovery program. Helps them share their stories, experiences as well as supporting each other and become sociable again step by step.

18 Courtyard area provides residents with an outside space to spend time outdoors and relax.

19 Variety of native planting and seasonal flowers provides natures elements as well as preventing trains commuters looking into the centre.

20 National Network Railway line running to and from Liverpool street and Bethnal Green.

21 Secret entrance by Fleet Hill Street tunnel. This entrance is used to bring in the victims into the centre without anyone seeing them.

SURVIVAL

Escaping from extremely dangerous, traumatic and fearful life...

Design aims to provide a secret, safe way out for young mothers and their children from Tower Hamlet Borough who currently experience domestic violence.

With nature as their protector and disguise, by camouflaging the centre's exterior and avoiding others discovering the building's real use, as well as providing them with healing properties and therapeutic activities.

Guided step by step these young mothers and their children work their way towards building a new healthy, happy life...

매우 위험하고, 충격적이며, 두려운 삶에서 탈출.

디자인의 목표는 현재 가정폭력을 경험하고 있는 젊은 엄마들과 그들의 아이들을 위해 비밀스럽고 안전한 탈출구를 제공하는 것이다.

센터의 외관을 위장하고 다른 사람들이 건물의 실제 용도를 발견하는 것을 피할 수 있도록 보호자처럼 그리고 변장이 가능하도록 도와주는 자연과 함께, 그들에게 치유 장소와 치료 활동을 제공하는 것이다.

안내된 이러한 단계에 맞추어서 젊은 엄마들과 그들의 아이들은 차근차근 새로운 건강하고 행복한 삶을 향해서 노력하고 있다.

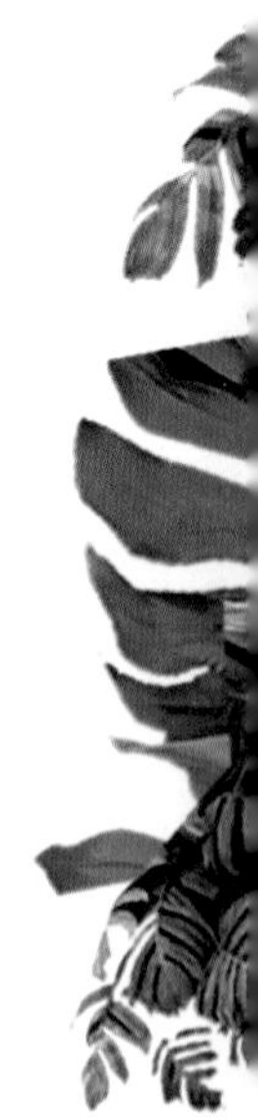

Wellness cafe and indoor garden

First stage shared accommodation

Front part elevation part section proposal

Back arches

POD DESIGN

1.Bishopsgate freight trains, used for transporting goods into the city.
2.Micro homes moving along railway track across the city, concept developed by students in Poland.
3.Moving houses on railway tracks by Swedish architects Jagnefalt Milton Andalsnes in Norway.
4.Pod inspirations began from the above precedences but also from the depot historical past, how was used, its connection to trains. Design was also influenced by the word "transient" which links back to the victims residents program in the centre, as they are homed in temporary accommodation in the back depot arches, where they share with other victims and then continues to the pod as next step to slowly gain the confidence to live by themselves and their child (whist still living within the centre). This will help them in the transition to confidently move out of the centre once recovered and begin a new life outside the centre.

Pod side elevation

Roof glass window to maximise natural lighting and contemplate stars at night.

Steam release pipe

Pod front elevation

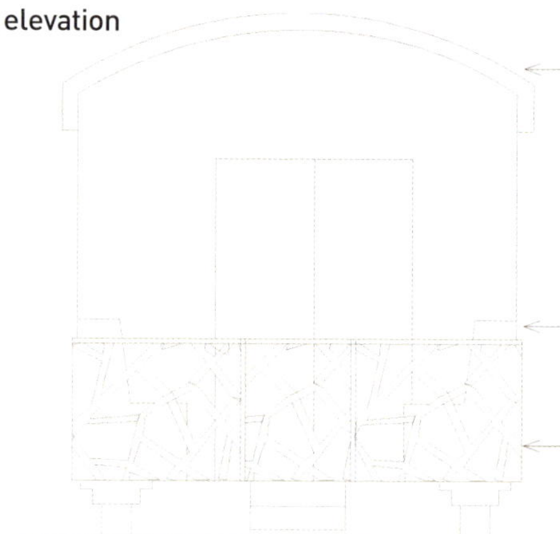

Extra roof to cover the balcony area.

Sitting bench for relaxation outside in balcony space.

Iron balustrade to protect from falling when in mini balcony

Pod section b-b with dimensions

Steam release pipe

Pull out table

Pod section a-a

Roof glass window

Fold-able beds from wall. This offers the opportunity to make the most of the interior space when the beds are not needed. Top bed accessed by separate ladder.

Storage for personal belongings.

Pod floor plan

Under table drawers to keep personal belongings.

Pull out table from under work top for when needed whilst cooking and eating, cupboards under table.

Pod includes kitchen steam system to mantain the wellness of cooking steam vegetables and others.

Sink for every day use and hidden storage spaces.

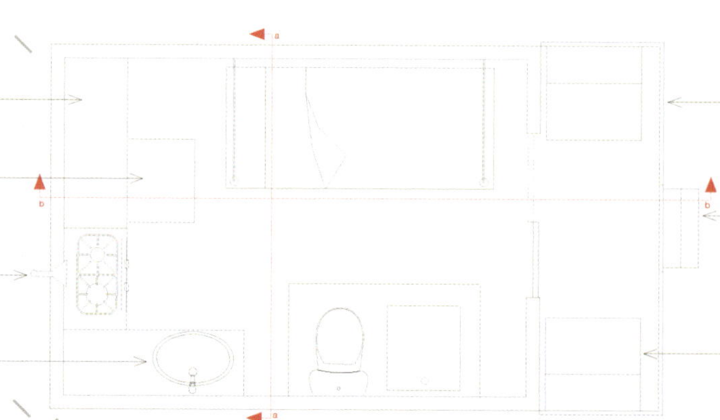

Pod includes a mini balcony to offer outside inside space for the occupant, surrounded by balustrade for protection.

Train inspired steps for descending from pod

Pod balcony includes sides sitting area to contemplate the view and enjoy time outside whilst still in the pod.

Pod entrance elevation

Pod inside depot before travelling up to the high-line

Pod side elevation

Pod travelling up to the high-line

Pod section view on high-line

Action Station

Lina Danileviciute
Interior Architecture and Design
Studio 2 NEXUS / Year 3

Studio Int 02: Action Stations:

Studio 2 will firstly explore the notion of Escapism by looking at the future of retail design, its response to both the new generation of immersive and physical spaces that allow the consumer to participate and generate meaningful and expressive spatial experiences.
Solving the displaced people issue, by designing their temporary home which helps them to heal and makes then capable to find new home and be in community.

스튜디오 2는 먼저 소비자들로 하여금 참여하고 의미 있는 표현적인 공간 경험을 만들 수 있도록 가능하게 해주는 생동감 있는 그리고 물리적인 공간의 반응을 보여주는 소매지역 디자인의 미래에 대한 관찰을 통해서 현실도피의 개념을 살펴 볼 것이다.
갈 곳을 잃은 사람들이 공동체의 일환이 되고 새로운 집을 찾을 수 있도록 능력을 가질 수 있도록 해줌과 동시에 그들을 치료할 수 있는 임시적인 집을 디자인 함으로써 이러한 문제에 대한 해결.

The Borough was formed in 1965. The borough lies to the east of the City of London and north of the River Thames. The borough has a population of 272,890,[2] which includes one of the highest Muslim populations in the country and has an established British Bangladeshi and Pakistani business and residential community.

자치구는 1965년에 형성되었고 런던시의 동쪽과 템즈 강의 북쪽에 있다.
이 자치구는 272,890명의 인구를 가지고 있는데, 이 나라에서 가장 높은 이슬람 인구 중 하나이며, 영국계 방글라데시와 파키스탄의 사업과 주거 공동체를 가지고 있다.

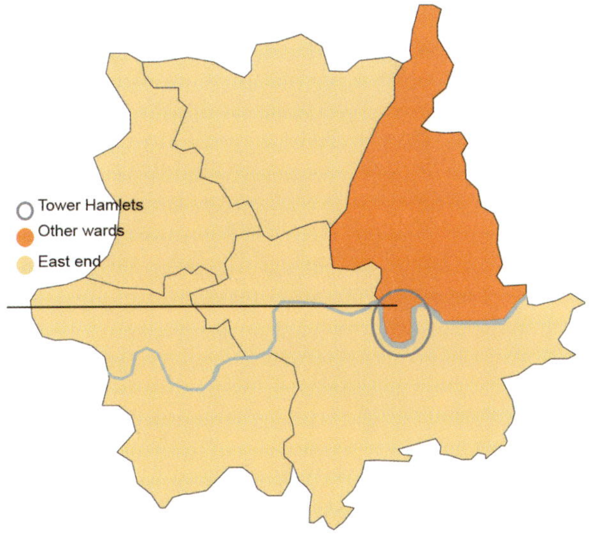

East End London map: its an area of Central, East London and London Docklands. Historically, the East End has always contained some of the poorest areas of London.

이스트엔드 런던 지도: 그곳은 중부, 동부 런던, 런던 도크랜드의 지역이고, 역사적으로 이스트 엔드는 항상 런던의 가장 가난한 지역들을 포함해왔다.

100 areas of parks and open spaces within its boundaries. Including biggest park Victoria Park: 86.18 hectares (213.0 acres), created 1884.

그 경계 안에 있는 100개의 공원과 열린 공간. 가장 큰 빅토리아 공원을 포함: 86.18헥타르(213.0에이커)가 1884년에 만들어졌다.

Tower Hamlets is a riverside borough, and one of the largest open spaces is the Thames itself.

타워 햄릿은 강변의 자치구이며, 이는 템즈강 자체로 볼 때 가장 큰 열린 공간 중 하나이다.

Tower Hamlets borough map: is a London Borough in East London which covers much of the traditional East End. The borough has a population of 272,890, which includes one of the highest Muslim populations in the country and has an established British Bangladeshi and Pakistani business and residential community. Tower Hamlets has the highest proportion of Muslims in England outnumbering Christians, and has more than 40 mosques and Islamic centers.

타워 햄릿 자치구 지도: 이스트 엔드의 많은 부분을 차지하고 있는 런던 자치구입니다. 이 자치구는 272,890명의 인구를 가지고 있는데, 이 지역은 이 나라에서 가장 높은 이슬람 인구 중 하나이며, 영국 방글라데시와 파키스탄의 사업과 주거 지역 사회를 가지고 있다. 타워 햄릿은 영국에서 기독교인보다 이슬람교도 비율이 가장 높고 40개가 넘는 회교 회관을 가지고 있다.

Fleet hill street tunnel, now a place to film movies and for photo-shoots.

Situated just off Brick Lane, Shoreditch tube station was the northern terminus of the East London Line.

Site front yard, windows been blocked with thermalite aircrete blocks

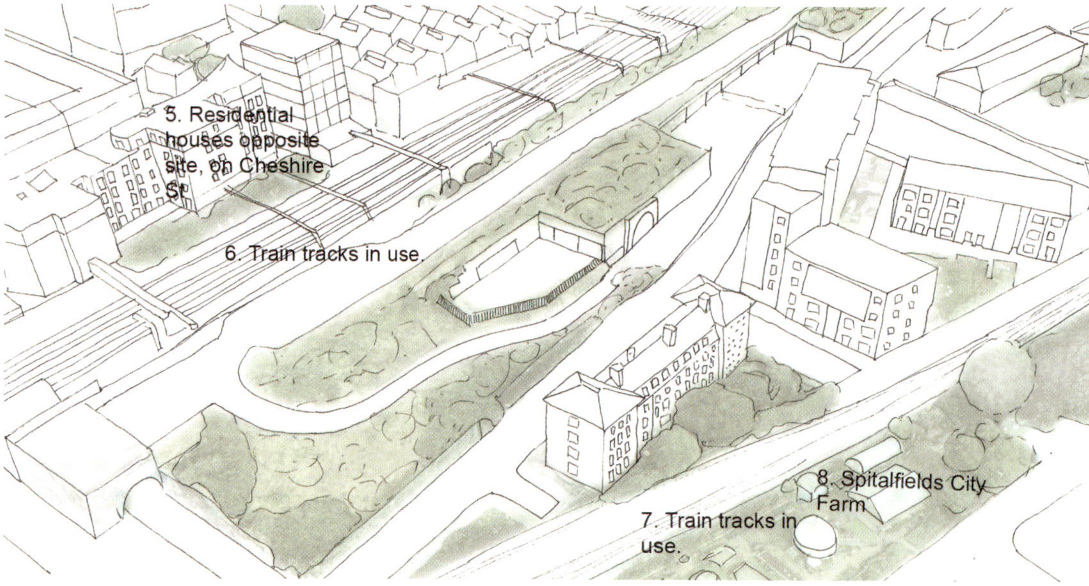

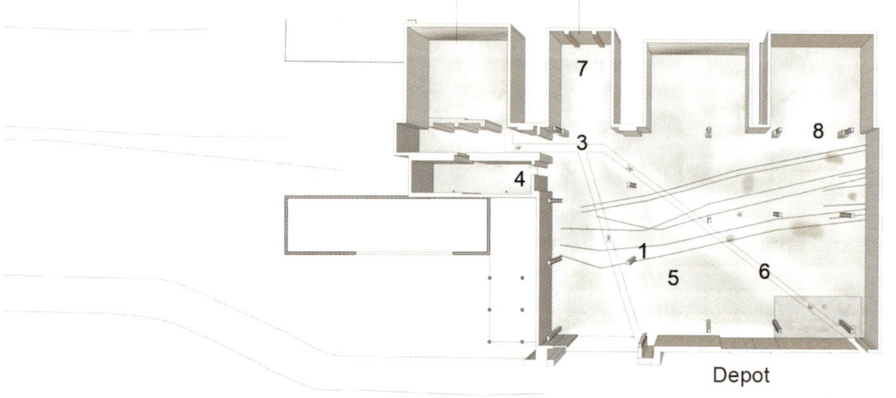

Space has 4 arches inside, every arch is different size and contains different materials inside.

Electricity point near exit. Every arch has sockets and florescent ceiling lights.

Concrete base for steel beam.

Possibly stairs to go down, but where blocked and unused.

Rails across the main area might be used for cattle transport from roof top to inside. Cracked concrete and cobble flooring with many holes and broken concrete.

Fire exit

Aluminum sheets covering back of 3 arches.

Old arch behind the metal sheet, possibly it was arched window.

The focus of the studio "Nexus" is on exploring displaced people emotions and needs, by providing them with temporary home, that will help them to stand up on their feet and start new period in their life.
The site is pedley street, a former goods yard, situated at the end of a closed road, adjacent to north east rail way line, nomedic garden and social houses, within the borough of Tower Hamlets.
The aim of the project is to provide a community space that raises money for displaced animals and displaces people community.
Through primary and secondary research the programme develops a series of bespoke spaces and habits that focuses on the animal needs as well as people needs.
Aim is to give this place and building identity, which will give positive atmosphere towards community and displaced people.
The programme creates private, observable and semi private space, along side community education space, through decreate and dynamic moves at utilising human and animal anthropometric and ergonomics.

스튜디오 "넥서스"의 초점은 그들이 일어서서 새로운 시대를 시작하는 데 도움을 줄 임시 집을 그들에게 제공함으로써 난민들의 감정과 필요를 탐구하는 데 있다. 이 장소는 폐쇄된 도로 끝에 있는 타워 햄릿 자치구 내의 북동 철도노선과 유목 정원, 사회 주택에 인접해 있는 이전의 물품 보관소인 페들리 거리이다. 이 프로젝트의 목적은 갈 곳을 잃은 동물 사람들을 위해 기금을 모으고 지역사회 공간을 제공하는 것이다.
1차 및 2차 연구를 통해 프로그래머는 사람들이 필요로 하는 것뿐만 아니라 동물의 필요에 초점을 맞춘 일련의 맞춤형 공간과 습관을 개발한다.
이 목적은 이 장소와 빌딩에 정체성을 주는 것인데 이것은 갈 곳을 잃은 사람들과 공동체에 대한 긍정적인 분위기를 전달할 수 있다.
프로그래머는 인간과 동물의 인체측정학 및 인간공학을 활용하여 장식과 동적인 움직임을 통해 지역사회 교육 공간과 함께 개인적이고 관찰 가능한 반개인적인 공간을 만든다.

"the dog house"

WHAT IS HOME
PRIMARY RESEARCH INTO FAMILY

Mother (Age 48)

Father (Age 47)

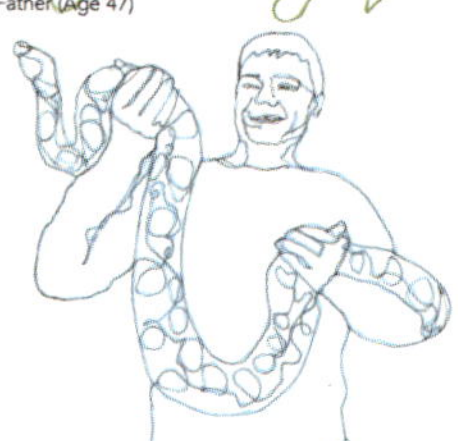

Son 1 (Age 24)

Son 2 (Age 20)

Son 3 (Age 18)

Son 4 (Age 16)

Son 5 (Age 15)

Daughter (Age 12)

Son1:
Home is where you can feel personal freedom, actions and thinking are expressed in different ways compare to the work place. Personally, home become home when your family is around and the uproar that family makes around my-self. In a first place I would miss my family members including Pets that was part of our family. Also, personal belongings such as pictures that reminds the back days when I was child. If I have to move the house I will take all my photo albums and the plants that I grow personally.

집은 여러분이 개인적인 자유를 느낄 수 있는 곳이고, 행동과 생각이 일하는 장소와 비교하여 다른 방식으로 표현된다. 개인적으로, 집은 당신의 가족이 곁에 있을 때 그리고 가족들이 나 자신을 둘러싸고 하는 시끌벅적함이 있을 때 비로소 집이 된다. 우선 우리 가족의 일부인 애완동물을 포함하여 가족들이 그리울 것이다. 또한, 내가 어렸을 때 기억나는 사진과 같은 개인적인 물건들도. 만약 내가 집을 옮긴다면, 나는 나의 모든 사진 앨범들과 개인적으로 키워진 식물들을 가져갈 것이다.

Son3:
Home is soothing I can go to and spend my time off doing what I like. What makes home? The things that you have such as furniture, technology and pets. I would miss my computer that's where I do all my work. To other place I would take my Aquarium, computer and furniture. I lost and I want to get back a few more of my gold fish and some furniture.

집은 내가 갈 수 있고 내가 좋아하는 일을 하면서 시간을 보낼 수 있는 위로가 있는 곳이다. 무엇이 집으로 만들까? 당신이 가지고 있는 가구, 기술 그리고 애완동물. 모든 작업을 수행하는 컴퓨터가 그리울 거예요. 다른 곳에 간다면 나는 수족관, 컴퓨터, 가구를 가지고 갈 것이다. 잃어버린 금붕어와 가구 몇 개를 더 사고 싶어요.

Son4:
Home is a cosy place where you dedicate most of my life living in – It is a place where you will return after being away and a place that will keep you safe. Usually at home, you would have someone waiting for you to get back – whether it is your partner or your family. I would miss the memories created at home – from birthdays to playing board games. I would miss the cosiness of my personal room and my comfort. I would miss my personal stuff such as the computer because I spend most of my current time on the computer learning and experiencing new things. To other place I would take all my belongings, from clothes to books. The clothes would bring a smell that would make feel like I belong there. I would bring my bed or a new bed that gives me comfort because without a comfortable bed there is no point returning when you can go and sleep anywhere else. I would bring decorations such as posters because they would make me feel part of the house as they have a personal connection towards what I enjoy. If I was able, I would bring back the apple tree that sits in the countryside of Lithuania because this tree was my source of food and enjoyment – being able to climb it and rest was elevating. In addition, I would bring back all the old and antique board games which created joy whenever it was needed because every board game had a unique style and playing one game after another was never the same as the last.

집은 당신이 삶의 대부분을 보내면서 머무는 아늑한 공간으로 한동안 떠난 뒤에 다시 돌아 가는 장소이면서 안전하게 당신을 보호할 수 있는 장소이다. 대개 집에 있는 경우 파트너든 가족이든 간에 누군가가 돌아오기를 기다리고 있는 곳이다. 생일부터 보드게임까지 집에서 만든 추억이 그리울 것이다. 나는 내 개인 방과 개인의 안식을 주는 편안함이 그리울 것이다. 난 컴퓨터 같은 나의 개인적인 것들을 그리워할 것인데 그 이유는 나의 현재 시간의 대부분을 컴퓨터를 배우고 새로운 것들을 경험하는데 쓰기 때문이다. 나는 옷에서 책까지 모든 소지품을 가지고 다른 곳으로 갈 것이다. 그 옷들은 마치 내가 거기 있는 것처럼 느껴질 냄새를 가져다 줄 것이다. 편안한 침대가 없다면 내가 가고 돌아오는 것이 아무 의미가 없기 때문에 나는 나를 편안하게 해 줄 내 침대나 새로운 침대를 가져 갈 것이다.
나는 포스터와 같은 장식품을 가져갈 것이다. 왜냐하면 포스터는 내가 좋아하는 것과 개인적인 관련이 있기 때문에 집의 일부를 느낄 수 있기 때문입니다. 만약 내가 할 수 있다면, 리투아니아의 시골에 있는 사과나무를 가져갈 것이다. 왜냐하면 그 사과 나무는 나의 음식과 즐거움의 원천이었기 때문입니다. 그 나무를 타고 올라서 쉴 수 있는 것이다. 게다가, 나는 그것이 필요할 때마다 즐거움을 만들 수 있는 오래되고 독특한 보드 게임을 가지고 갈 것이다. 왜냐하면 한 게임씩 차례로 하는 것이 결코 이전과 같지 않기 때문이다.

KEY POINTS FROM FAMILY

Garden - Flowers - Burning candles - Sewing - Smell - Squirrels - Technology

Dinner - Wooden table - Slippers - Technology

Freedom - Pets - Memories - Plants

Noise - Smell - Memories - Guitar

Pets - Technology - Aquarium - Fishes

Heat - Storage - Memories - Board games - Comfort - Technology - Smell - Apple tree

Technology - Noise - Toy

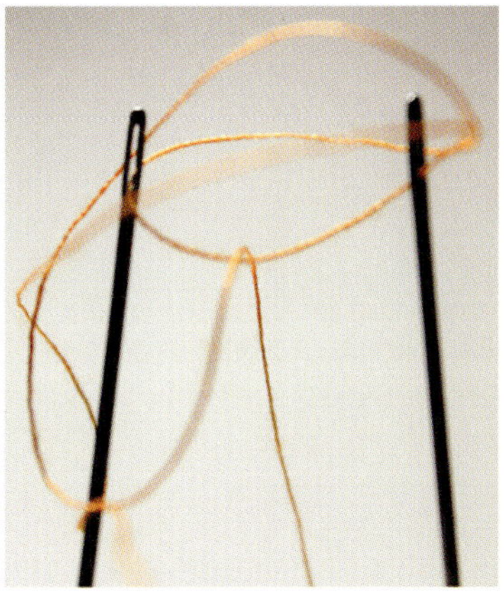

Son 4 (Age 16)

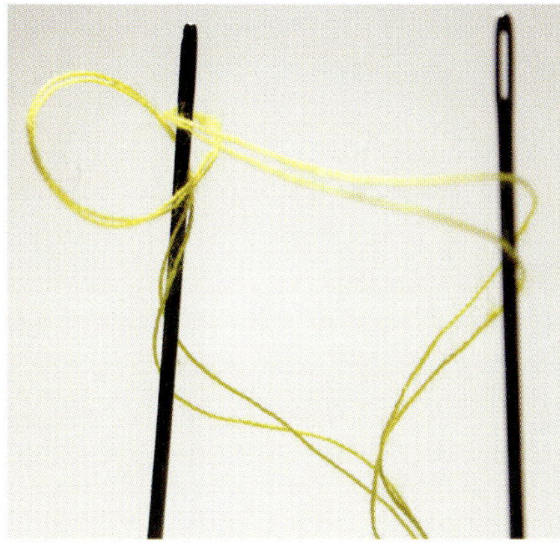

Mother (Age 48)

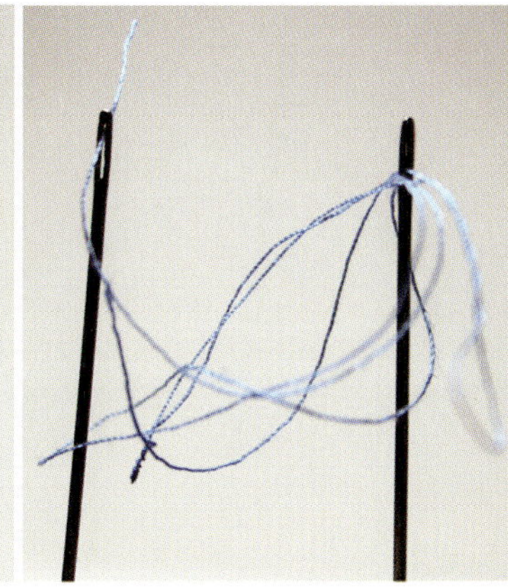

Son 1 (Age 24)

Son 2 (Age 20)

1. Stairs/Bathroom. Mum and dad just comes upstairs to have a shower, or wake someone up for work or school, but this happens not often.

2. All children stays in their rooms most of the time, its just when someone needs something they go in each other rooms, just boys do not go so often to sisters room as she does.

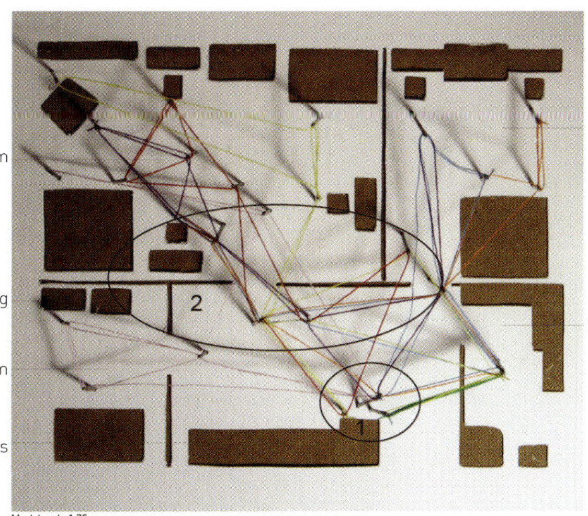

Model scale 1:75

1: Garden, where mother spends a lot of time growing flowers and vegetables. Also its morning coffee's main exploration point, where she explores animals that visit their garden, as well as green areas. That calms her down and "cleans" her mind.

2: Kitchen-corridor and stairs to upstairs is most common space where all family face each other. Meeting point.

3:Living room with parent room. Mother is the only one that spends her 90% of time in this space. She looks after her oldest daughter's child while parents work. So she is very active in his area and kitchen. Also if the child is not there she still spends most of the time her. Doing varies things from eating, drinking coffee (which is calm and relaxing time for her) sewing, doing varies DIY's.

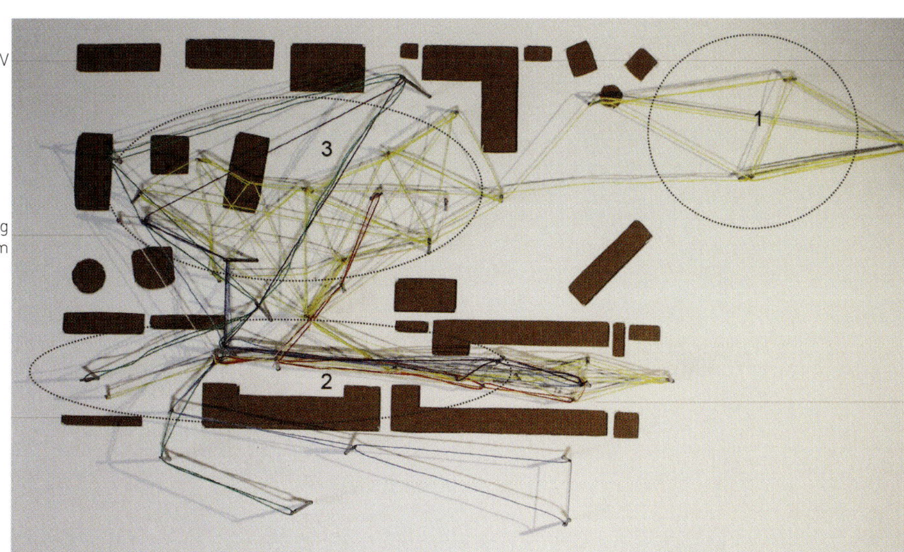

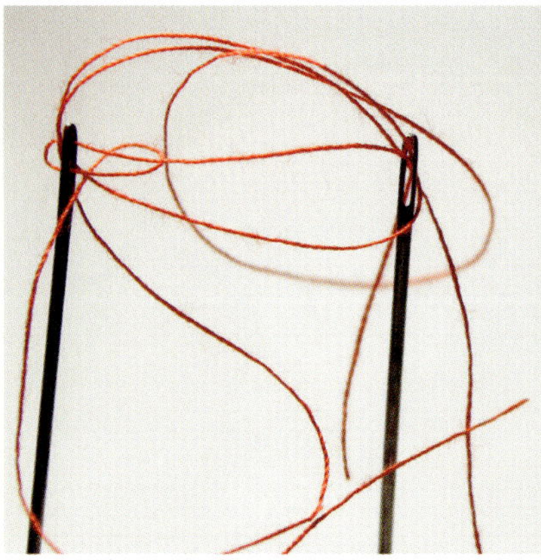

Son 5 (Age 15)

Daughter (Age 12)

Son 3 (Age 18)

Father (Age 47)

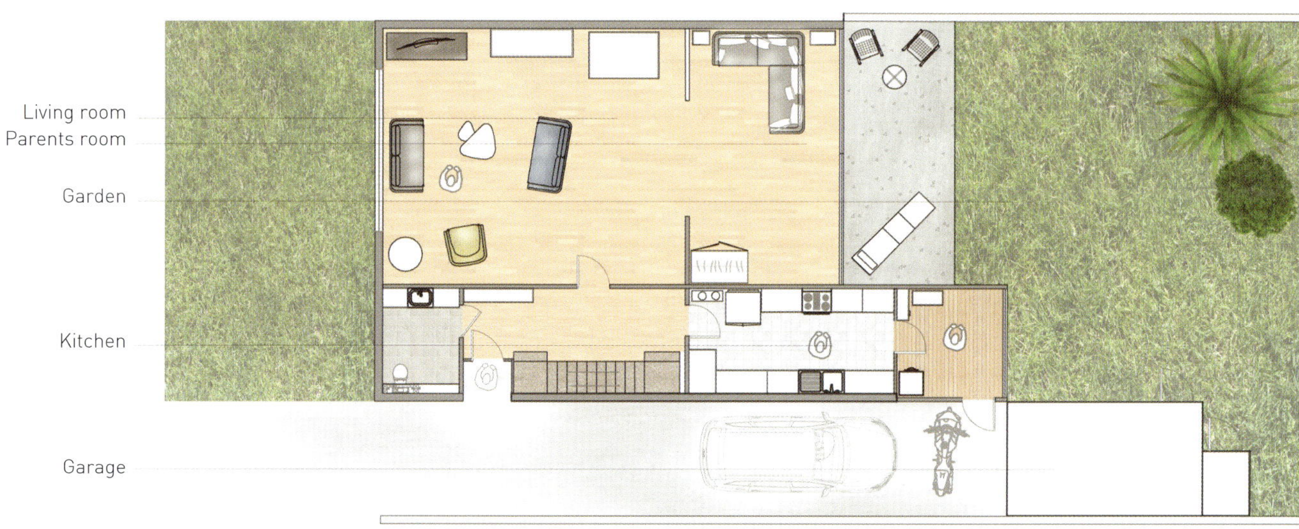

Stroking animal:

-Stroking a living creature, whether hard-shelled or furry, relieved anxiety.

Horse:

-Activities like grooming a horse and leading one around a pen have been shown to reduce PTSD symptoms in children and adolescents.

Fish:

-Dined in front of aquariums with brightly colored fish, they ate more, got better nutrition and were less prone to pacing. They were also more attentive and less lethargic.

Guinea pig:

-A guinea pig in the classroom, they were more social with their peers, smiled and laughed more, and showed fewer signs of stress.

Caring for a dog can help you make healthy lifestyle changes by:

- Increasing exercise.
- Providing companionship.
- Helping you meet new people.
- Reducing anxiety.
- Adding structure and routine to your day.
- Providing sensory stress relief.
- Staying connected.
- Dogs especially are great at encouraging owners to get exercise, and this can be beneficial for those suffering from depression.
- Dog will look into your eyes to gauge your emotional state and try to understand what you're thinking and feeling
- Dog owners are less likely to suffer from depression than those without pets.
- People with dogs have lower blood pressure in stressful situations than those without pets.
- When people with borderline hypertension adopted dogs from a shelter, their blood pressure declined significantly within five months.
- Playing with a dog or cat can elevate levels of serotonin and dopamine, which calm and relax.
- Pet owners have lower triglyceride and cholesterol levels (indicators of heart disease) than those without pets.

Heart attack patients with dogs survive longer than those without.

- Pet owners over age 65 make 30 percent fewer visits to their doctors than those without pets.

Having a dog or other animal that needs walk and taking care off you will exercise, and communicate with other people. Build confidence and release stress and lower blood pressure. Also there are all high-line where I can design spaces for people to plant and take care of their planting. Gardening is an opportunity for everyone to experience this kind of regular contact with nature, even if they live in built-up areas.

-Any type of gardening, whether it is in a home or allotment garden, is an opportunity for physical activity. Gardening is typically seen as moderate intensity exercise equivalent to playing doubles tennis or walking at a speed of 3.5mph, and so carries similar fitness benefits.

-Gardening is also linked to better diets. Home and allotment gardens have long been important for domestic food production, but gardening can also encourage people to eat more healthily and act as an educational resource on nutritious food.

-Research has shown that gardeners generally have greater life satisfaction, enhanced self-esteem and fewer feelings of depression and fatigue than non-gardeners.

So having a shelter for displaced animals and allowing displaced people come with their animals and live temporary will help them heal. People that don't have animals or lost them through their displacement shelter would provide possibility to take care, walk dogs even take them into their pods up on high-line and spend some time. This will help them to get better and also animals will get trust again for people, and will heal together with people.

당신은 애완동물을 데리고 걷거나 그것을 돌볼 것이고 다른 사람들과 의사소통을 할 것이고 이러한 것은 자신감을 키우고 스트레스에서의 해방과 좀 더 낮은 혈압을 유지할 것이다. 또한 사람들이 심을 수 있는 공간을 설계하고 식물을 가꾸는 데도 사용할 수 있는 하이라인도 있습니다. 비록 사람들이 만들어진 공간에 살더라도 정원을 가꾸는 것은 모든 사람들이 자연과 정기적으로 접촉하는 것을 경험할 수 있는 기회이다

집이든 할당정원에 있든 간에 이루어지는 정원 가꾸기는 신체활동을 위한 기회이다. 정원 가꾸기는 보통 테니스를 두 번 치거나 3.5mph의 속도로 걷는 것과 같은 적당한 강도 운동으로 보여지며, 따라서 비슷한 건강상의 이점도 있다.

정원을 가꾸는 것은 또한 더 나은 식단과 관련이 있다. 가정과 할당 정원은 오랫동안 지역 음식 생산에 중요했고, 정원 가꾸기는 또한 사람들로 하여금 건강에 더 좋은 음식을 먹고 영양가 있는 음식에 대한 교육 자원으로 활동하도록 장려할 수 있다.

연구에 따르면 정원사는 일반적으로 비 정원사들보다 삶의 만족도가 더 높고 자부심이 높으며 우울증과 피로감이 덜하다고 한다.

그래서 갈 곳을 잃은 동물들을 위한 은신처를 마련하고, 갈 곳을 잃은 난민들이 동물들과 함께 오는 것을 허락하고, 일시적인 생활을 하는 것은 그들이 치유하는 것을 도울 것이다. 동물이 없거나 그들의 이주에 의한 동물을 잃어버린 사람들은 하이라인에서 산책하는 개들을 그들의 Pod에 데려가서 시간을 함께 보낼 수 있다. 이것은 그들이 나아지는데 도움을 줄 것이고 또한 동물들은 다시 사람에 대한 신뢰를 가질 수 있고 사람들과 함께 치유될 것이다.

-"A pet can remind you that you're not alone," says Desiree Wiercyski, a life coach in Fort Wayne, IN.

"Pets offer unconditional love, which can be extraordinarily soothing when feeling isolated."

-Just by stroking, sitting next to or playing with a pet can gives owning a chance to relax and calm their minds.

GUIDELINES FOR THE DESIGN AND MANAGEMENT OF ANIMAL SHELTERS BY RSPCA

Battersea Dogs & Cats Home daily routine:

Get into work by 8am and mentally prepare for the day.

The first job is the morning clean. Cleaning is something will take 80% of your time doing.

Then get started on the dogs' cats breakfasts. Every block has an enrichment chart which workers can consult for ideas on how to feed the dogs that day so that they can have a little fun with their mealtimes.

For example, use toys, peanut butter and frozen food to make things more exciting.

Later in the morning they get to work on walking the dogs. We take them around the site,using their secure paddocks so they can run around off their lead; and also take them to Battersea Park. They intend for dogs to spend as little time here before they get their forever homes, it is good to try and build a good base for them behaving at home.

Time to update computer system with notes about each of the dogs so that everyone who works with them is kept fully updated.

After I have my own lunch, it is time to focus on the dog's lunches.

As the day comes to an end, we get to work on making the dogs comfortable for the night. We ensure that all the yards and kennels are clean, and that the dogs all have nice warm blankets for bedtime. We take them out for another walk so that they can go to the toilet.

When it's my home time, they already looking forward to the next day when they can see the dogs again.

PROPOSED FLOOR PLAN

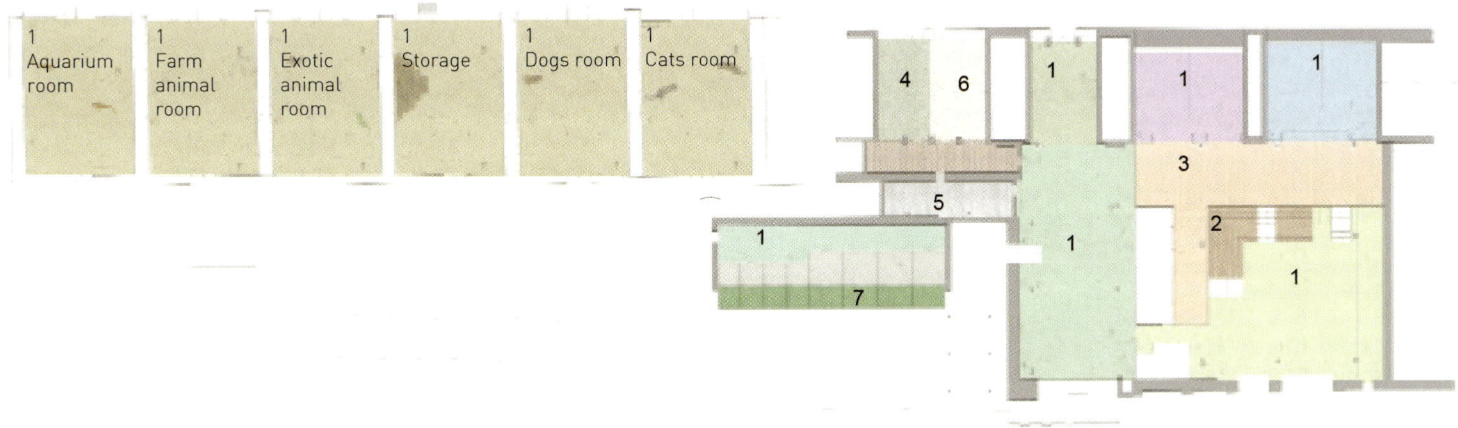

1.
Hanson concrete can incorporate a proportion of Hanson Regen, (Ground Granulated Blast furnace Slag or 'GGBS'), which is a cement replacement material, helping to produce a product that generates a much lower level of CO2 emissions than ordinary cement.
Colourful concrete in most of the rooms is to produces better day for people who works there all day every day. To increase their wellbeing (colours choice according to The Effect of the Work Environment on Workplace Wellbeing – Research at a Youth Centre by Fehér, Hanna).

2.
Walnut wood stairs:
American black walnut. Strong and stable, this top quality hardwood offers a mostly straight grain and ranges from dark chocolate to a pale brown in colour. It is versatile and extremely hard-wearing.

3.
Ranch oak flooring:
Ranch Oak is a warm and natural wood floor with rustic charm. Ranch Oak is treated with our new, extra matt lacquer which combines the performance of low-mainnance lacquer with the unique look of untreated wood.

4.
Wet room flooring:
Tarasafe Ultra H2O is a High performance slip resistant safety flooring for wet and dry, barefoot and shod applications. This enhanced slip resistant safety vinyl floor covering has a raised emboss designed for applications where there are continuous wet areas. Its is also treated with an anti-bacterial and fungicidal treatment, has a 12 year warranty and is 100% recyclable. (Colour choice based on The Effect of the Work Environment on Workplace Wellbeing – Research at a Youth Centre by Fehér, Hanna)

5.
Kitchen flooring:
A light, unfinished concrete effect floor by Japanese architect Tadao Ando. These large vinyl tiles create the appearance of contemporary raw concrete slabs, but are far more durable and warmer underfoot than a real concrete floor. Environmentally Friendly. Tough & Durable.

6.
WC flooring:
Mardi Gras 91 Modus vinyl flooring. The light core colouring provides the ultimate contrast for the grey and black fleck that feature, adding to the appearance of texture. Easy to clean and water resistant. Slip resistant.

7.
Artificial grass for dogs:
Bahamas is a highly realistic artificial grass, featuring 3 colour tones of dark green, muted olive and subtle caramel thatch, in a gorgeous imitation of natural grass. Doesn't contain any lead or cadmium. Easy to clean and pet friendly. This artificial lawns are ideal for all pets and can be easily cleaned using a garden hose, warm water and mild detergent, or even with a good rainfall.

RSPCA ANIMAL SPACE:
1. FLOORS: To guard against infection, all floors should have a smooth impervious surface e.q. tiles. Concrete can be used if it is hardened and treated to be impervious. In outside areas, gravel flooring is preferable to grass or earth, but must be thoroughly washed frequently. Wood is not advised as it deteriorates and paving is also difficult to keep clean.

1. 바닥: 감염을 방지하기 위해 모든 층은 타일과 같은 매끄러운 불침투성 표면을 가져야 한다. 콘크리트는 경화되고 불침투성이 되는 경우 사용할 수 있다. 외부 지역에서는 자갈 바닥이 풀이나 흙보다 선호되지만 자주 철저히 씻혀져야 한다. 나무는 퇴화되고 포장도 청결을 유지하기 어렵기 때문에 적절하지 않다.

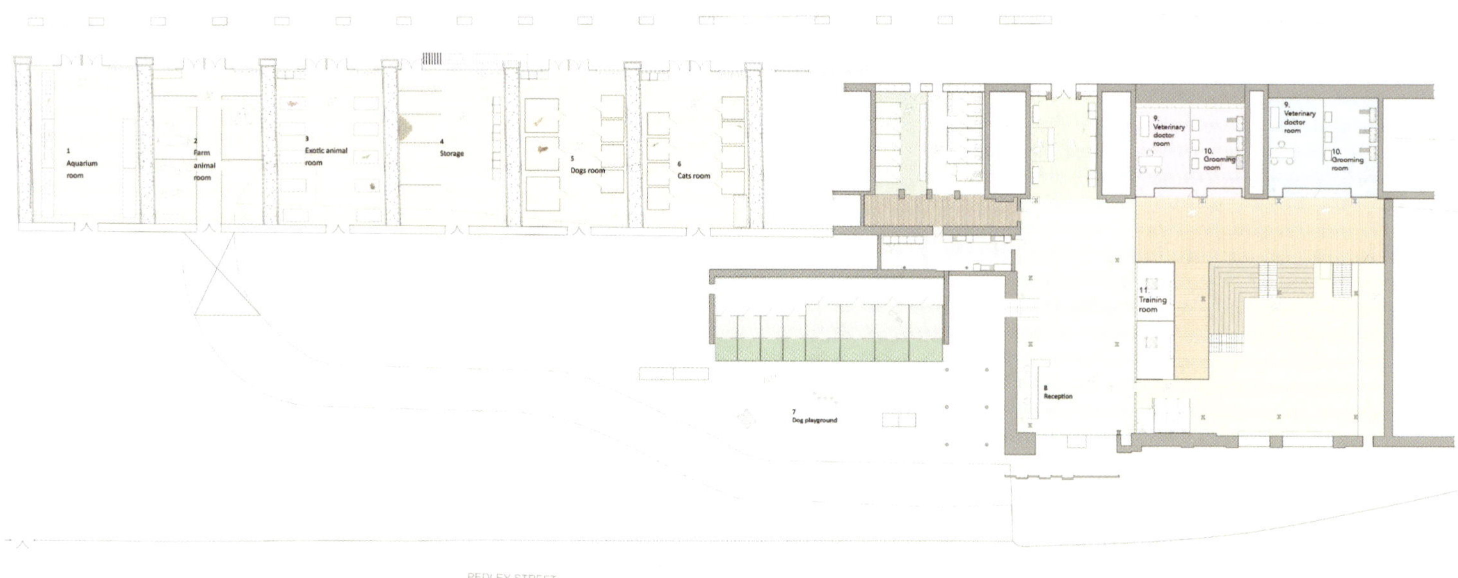

1. Aquarium room requires different size aquariums with storage space and water access.

Supplier swelluk

2. Farm animal room needs water access and wooden barriers between them. Allowing animals to see through fence will help them communicate with other animals.

Supplier wickes

3. Exotic animals needs heat lighting, heat pad, vivariums, water supply, thermometer, storage space and fridge to keep food frozen.

4. Storage needs for all food and other animals supplies. Hay, dray and wet food, beddings etc...

Supplier furniture-work

5. Dog room needs fridges for their food, water supply. Every dog needs to be separated as this room will hold dogs with health problems or other issues.

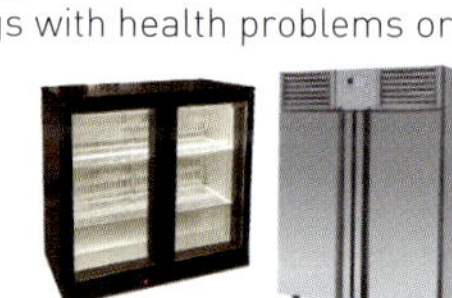

Supplier eco-fridge, all fridges and freezers through all space are used the same

6. Cat room will hold max 8 cats at once, needs food storage fridges and cupboards as well as water supply. Cat toys and accessories are essential for them to be busy and get healthier faster.

Supplier monsterpetsupplies

7. Dog playground includes a lot of agility training equipment for dogs to get better and exercise more, this will help them with communicate with people.

Supplier doggyjumps.

8. Reception for veterinary doctor, grooming room and educational programs.

Supplier officereality, gonna special made a reception table in this shape.

9. Veterinary doctor room, needs examination table, storage and work table.

Supplier technik-veterinary

10. Grooming room needs grooming table, bath, storage.

Supplier groom-ers-online Supplier simpsons

11. Training private room

Supplier
Livin The Dog Life – Gyms For Dogs

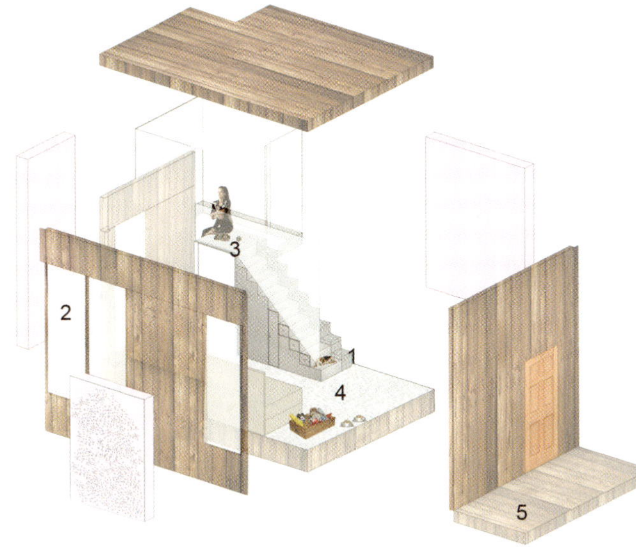

1. Bee house, for encouraging bees population
2. Cross laminated timber
3. Observation nest, from primary research of two families, seeing nature and having a calm space to relax helps them to relax and feel better.

1. Stairs for dogs, reflecting primary research of families.
2. Big windows to help people to see more nature while they are recovering. Also primary research of families showed that people sleep with open curtains and that helps them to wake up much happier and more positive.

3. Allowing pets to go on stairs, spending time together helps to get better quicker, as my research outcome showed.
4. Soft materials will provide people and animals with comfort and feeling warm and safe.
5. Encouraging people to go outside and stay in nature more.

POD HOUSE

1. Grass
2. Observation nest
3. Cross laminated timber
4. Animal friendly planting
5. Moss

Washable surfaces must be used to clean dog houses to avoid any bacteria that might cause various illnesses for animals and people.

1. Natural light, fresh air
2. DOG ARTIFICIAL GRASS
latex free artificial grass for dogs
3. TV screen, showing different movements for dogs inspired by Dominic Wilcox designs contemporary art exhibition for dogs.
4. Inside space
5. Bright colours for dogs because they can recognize colour levels, but more importantly its for people working there.
6. Stairs reflects primary research from one of the families, and how dog is very interested in stairs and sleeping on landing.

7. Places for dogs to hide when they feel like they want, playing.
8. Every staff has App that can connect with any dog they want and see what they are doing, they can give a dog a treat and the mechanism will shoot it. Also if the dog brings ball in hole it will shoot it back to dog, so they can be busy when staff is not around them.

1. Mechanics garage
2. Fire exit corridor
3. Observation "Nest" for educational programs. In arches there are veterinary doctor and grooming rooms, where people can come and learn how to treat animals and look after them.
4. Educational space for people and schools to learn about different animals, how to take care of them and teach people how animals can change the wellbeing of the person.
5. Bringing animal to show people differences between each species, explaining what animals require by adopting them or buying, so people know if they can afford and will they be able to take care of certain animal.
6. Floor finishes in most of the spaces needs to be sealed very well because there will be a lot of animals in spaces, such as going to veterinary, grooming or educational programs.
7. Displaced people pods has "Nests" for people to observe and relax.

8. Bee house purple/blue colour to attract bees
9. Main entrance, reception for veterinary and grooming
10. Dog houses
11. Playground for dogs
12. Animal friendly plants, these have been chosen purposely for the welfare of the animals.

13. Providing shadow for dogs while they run in yard
14. Natural light:
RSPCA:
• Ensure all animals have as much natural light as possible.
• It is important for animals to experience day and night for their sense of well being and routine.
• All animals should be able to sleep peacefully at night without harsh lighting.

ELEVATION
ENCOURAGING WILDLIFE

2. Bee/bug boxes fixed on walls to attract as much as possible bees and encourage population.
3. Used blocked existing windows to fix bee/bug boxes for increasing population for natural habitat.

3. Used blocked existing windows to fix bee/bug boxes for increasing population for natural habitat.

4. Windows with bee/bug boxes for educational program, people can see from inside building whats going on inside the boxes and how insects lives, exploring natural habitat. Boxes can be replaced when needed.

They are within single double glazed units and the boxes are fixed to frames and they can be removed whenever needed.

5. Fire exit

6. Opened blocked window to let as much natural light as possible.

7. Kept existing crane bird art, because it added identity fitting with my concept.

8. Bee boxes on high-line surrounded by bee friendly flowers that provides food and safe places for them to populate.

Ratchaneewan Daengbunma
Interior Architecture and Design
Studio 2 NEXUS / Year 3

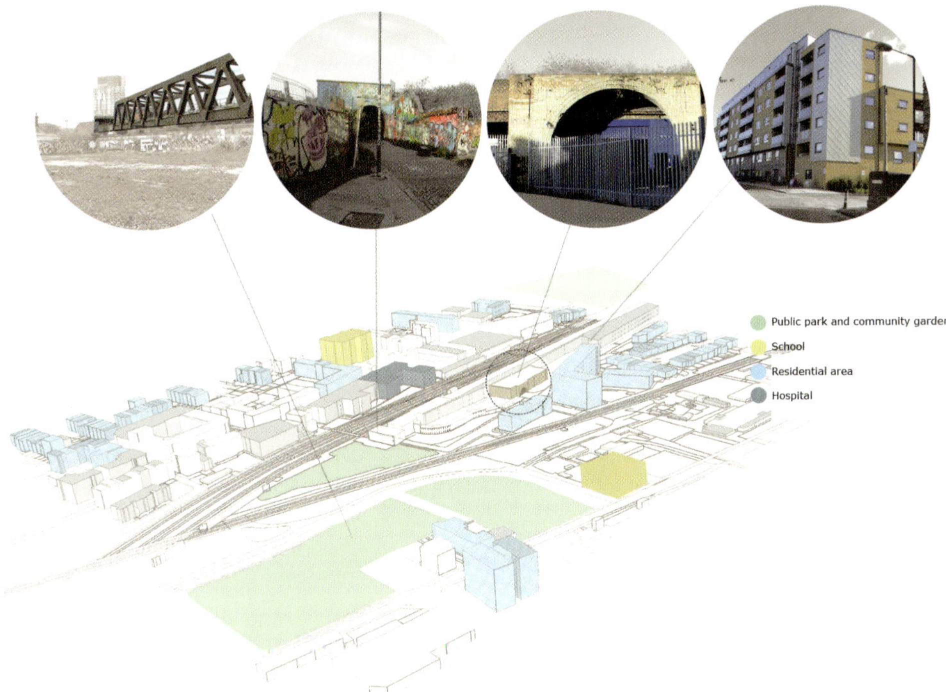

Site Location:
Pedley Street, London
E1 5ES

Pedley Street depot, is part of the Bishopsgate Depot and on the East London line. The depot was closed some years ago but mainly dealt with coal in its latter years. The site is now becoming run down and derelict, although its internal arches are rented to vehicle mechanics and scrap dealers, which are located between Brick Lane, Shoreditch and Bethnal Green.

There is only one road to access the site but Pedley street is not a busy road although it is close to the residential area. Pedley Street is in a conservation area so all external proposals will need careful consideration to be in keeping with the surrounding area and its character.

The site investigation revealed that the site and surrounding area is also derelict and neglected. There are the relicts of the previous working industry, monuments and ornate stonework at the south facing area. There is a lot of graf ti which makes the site feel very hostile but you can see the potential for developing the south facing outside area into a garden or green area and revitalizing the very characterful charm of the site.

페들리 스트리트 창고는 Bishopsgate 창고의 일부이며 런던 동부 라인에 있다. 이 창고는 몇 년 전에 문을 닫았으나 그 이후에 주로 석탄을 취급했다. Brick Lane, Shoreditch and Bethnal Green.사이에 위치한 차량 정비사와 고철 딜러에게 내부 아치가 임대되고 있음에도 불구하고 현재 창고부지는 해체되고 폐기되고 있습니다.

오로지 페들리 거리를 통해서 이 부지에 접근할 수 있는데 주거지역이 가까운 곳에 있음에도 바쁜 도로는 아니다. 페들리 거리는 보호구역에 있기 때문에 모든 외부 제안들은 주변지역과 그 성격에 부합하기 위해 세심한 고려가 필요할 것이다.

현장조사에서 알 수 있듯이 그 장소와 주변 지역 역시 방치되고 있다는 것을 보여주었다. 남향 지역에는 이전 작업 산업, 기념비, 화려한 석조 공사의 유물이 남겨져 있다.매우 적대적인 느낌이 드는 낙서가 많이 있지만, 남쪽을 바라보는 지역을 정원이나 녹색 지역으로 발전시키고 그 사이트의 매우 독특한 매력을 되살릴 수 있는 가능성을 볼 수 있다.

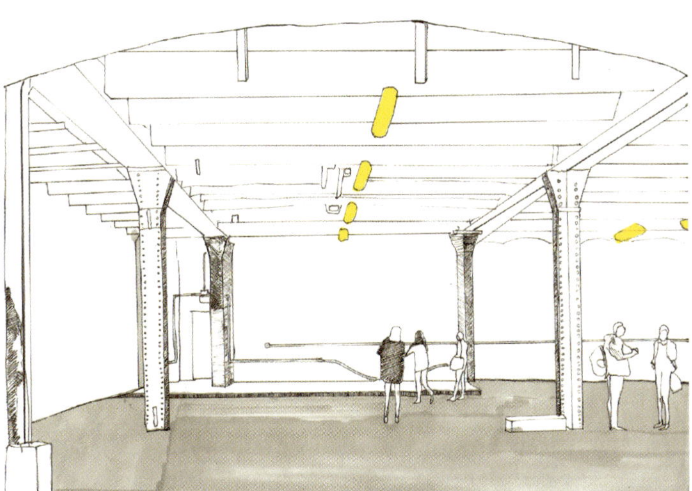

Survey sketch: View from arch 4

EXISTING HIGH LINE

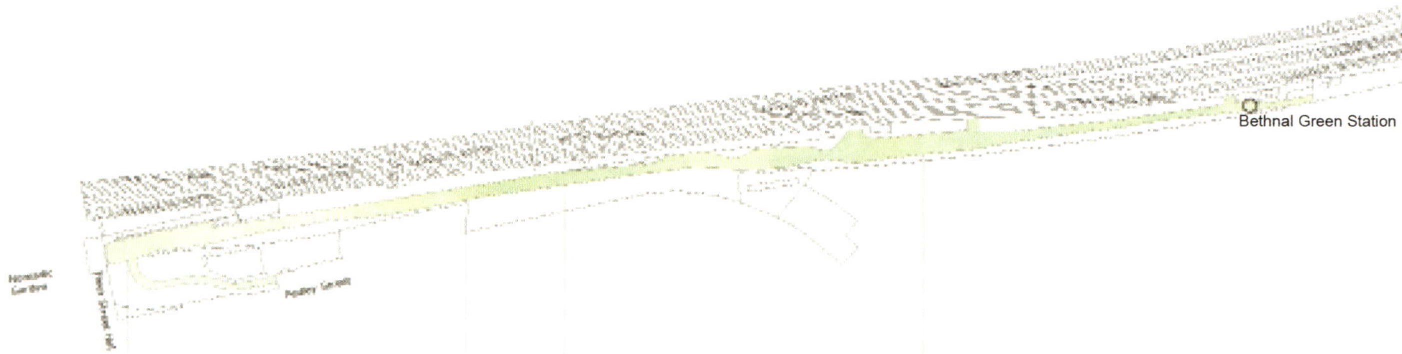

The high line runs all the way to Bethnal Green station. Looking south, the view is very impressive seeing all the way from Canary Wharf to the Gherkin. The elevated level receives unobstructed light from due aspect.

하이라인은 베스날 그린 역까지 쭉 뻗어 있다. 남쪽으로 보면 카나리아 부두에서 게르킨까지 이어지는 경치가 매우 인상적입니다. 상승된 레벨은 방향에 의한 한계 내에서 방해 받지 않는 빛을 수신합니다.

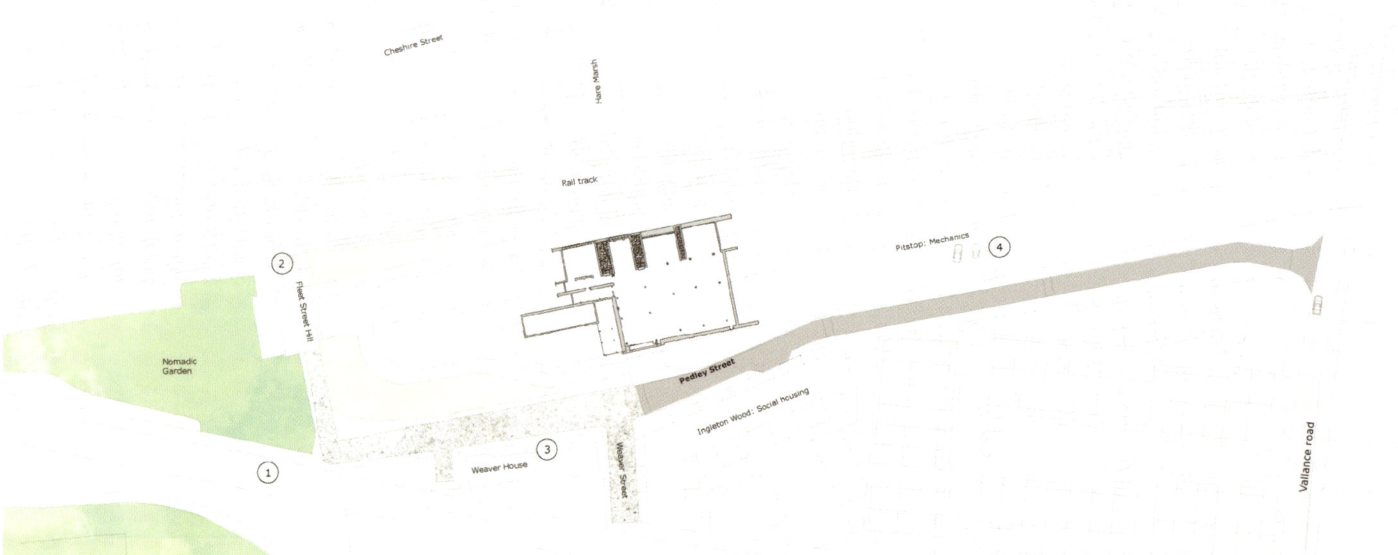

1. Allen garden has been cut through by the new railway structure.

2. Fleet Street Hill bridge to Cheshire Street

3. Weaver House: Residential dwelling

4. Part of the existing railway bridge, with the remaining structures protruding

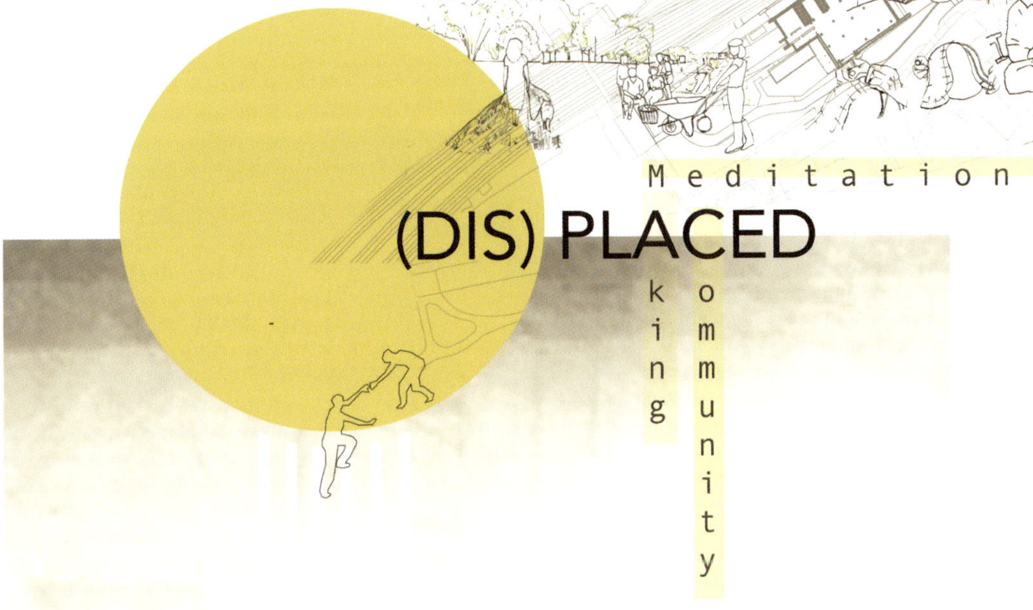

(DIS) PLACED

Making, Meditation and Natural Living

"Linking community through making and mindful living, incorporating nature and sustainable principles"

Provide support and homes for Displaced people.

The design proposal for the Pedley Street arches is to provide an area for people that have been displaced from their homes and communities.
People that have had to leave their home through natural or man- made events or disasters. Such as those affected by the Grenfell Tower fire.
The experience can be very traumatic and dealing with the challenges ahead can lead to physical and mental health issues.

페들리 거리 아치들의 디자인 제안은 그들의 집과 지역사회로부터 갈 곳을 잃은 사람들을 위한 공간을 제공하는 것이다. 자연적이거나 인재에 의한 사건이나 재앙을 통해 집을 떠나야 했던 사람들. 그렌펠 타워 화재의 피해자들과 같은 그 경험은 매우 충격적일 수 있고 피해자들이 이러한 경험을 이겨내기 위한 앞에 놓인 도전은 신체적, 정신적 건강 문제로 이어질 수 있다.

This is a space for people to bring together their creativity through arts, craft and making. Its an area for creativity, communication and contemplation and meditation. The idea is to support people that have gone through varying challenges by being displaced from their homes and communities.

이 공간은 예술, 공예, 제작을 통해 사람들이 그들의 창조성을 한데 모을 수 있는 공간이다.이 지역은 창조성, 의사소통, 사고와 명상을 위한 영역이다.기본적인 아이디어는 집과 지역사회로부터 갈 곳을 잃으면서 다양한 도전을 겪은 사람들을 지원하는 것이다.

PROPOSAL PLAN

HIGH LINE

1.Allotment
2.Garden Storage
3.Ramp for disabled access to high line
4.Network rail access road
5.Meditation walkway
6.Water feature
7.Sitting area
8.Compost

FLOOR PLAN

1.Storage
2.Temporary Accommodation Unit 1-10
3.Temporary Accommodation Unit 11-12: Disabled
4.persons
5.Meditation room
6.Fire Exit
7.Healing Garden
8.Weaving workshop
9.Sewing workshop
10.Storage
11.Exhibition area
12.Up-cycle workshop
13.Cafe
14.Coomunal kitchen
15.WC
15.WC
16.Storage
17.Fire Exit
18.Allotment

The allotment is south facing which provides good growing condition for many variety vegetables, fruit and flowers.
Allotment is available for the residents and also for volunteers.

시민농장은 남향이어서 다양한 야채, 과일, 꽃에 좋은 성장 조건을 제공한다.
시민농장은 거주자들과 또한 지원자들에 의해 이용 가능하다.

Meditation walkway and sitting area

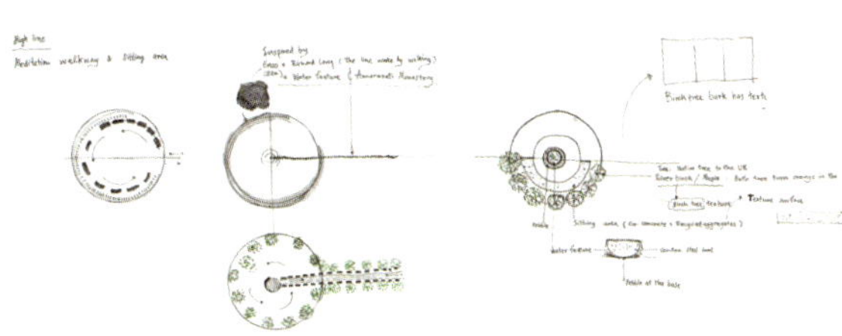

Meditation walkway and sitting area

Weaving workshop
Glazed wall walkway
Exhibition area: F

High line
COMMUNITY
patform
Community Cafe
Acoustic seating pods

Rail track
1
2
3
4
5
6
7
8
9
10
11
12
13
14
15
16
17
18
Entrance
Entrance
Section A-A
N

WALKWAY

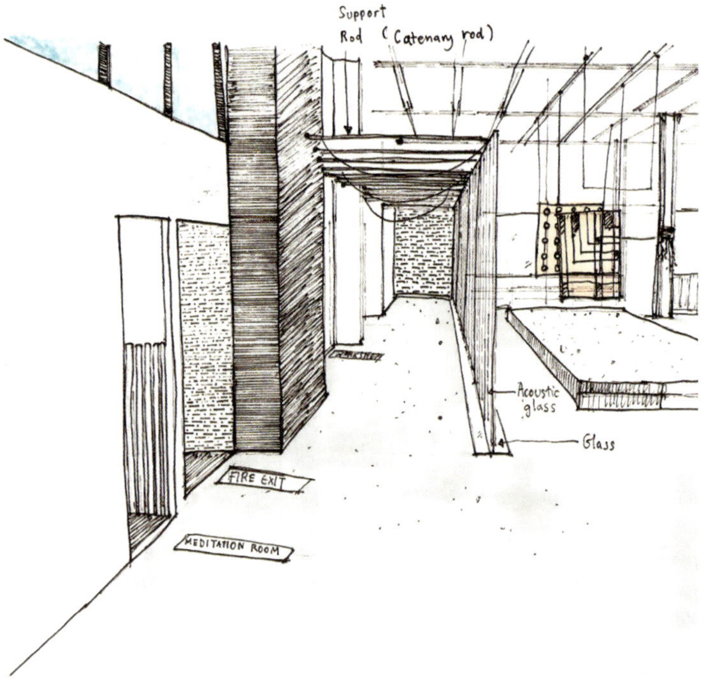

The walkway is inspired by Richard Long's work (A line made by walking).
Using the mirror strips in the floor to guide you through the space. It also adds light reflection and mirrors different part of the building as you walk through.
The supporting metal rods that secure the glazed wall can are use to drape fabric over them. this created the reflection of the existing jack arches.
Calm colour: For the walkway and exhibition area, a shade of pink has been used for the glazed partition to create a calming atmosphere, and an interesting interaction with the natural light that is being reflected around the space.

산책로는 리차드 롱의 작품(걸어서 만든 선)에서 영감을 얻었다. 바닥의 거울 스트립을 사용하여 공간 안내하며 그것은 또한 빛 반사를 더하고 여러분이 걸을 때 건물의 다른 부분을 거울처럼 비추어준다.
유리 벽을 고정하는 지지 금속 막대는 그 위에 천을 덮는 데 사용될 수 있다.
이로 인해 기존 잭 아치가 반사된다.
차분한 색상: 통로 및 전시장의 경우, 차분한 분위기를 만들기 위해 유리를 끼운 칸막이에는 분홍색 그늘이 사용되었으며, 공간에 반사되는 자연광과의 흥미로운 상호작용이 있었다.

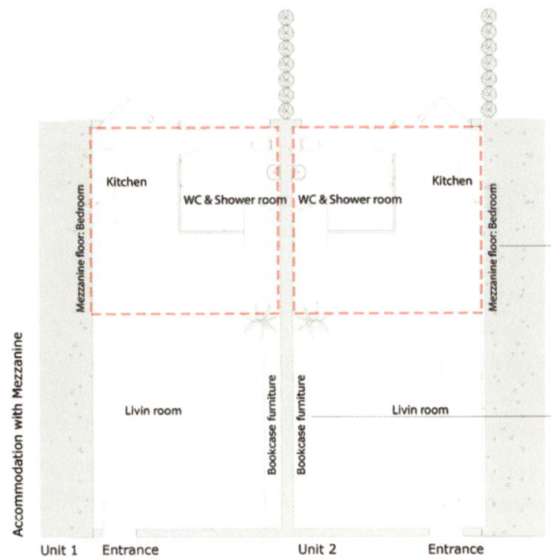

The bookcase furniture is a multi-use area for storage and sitting.

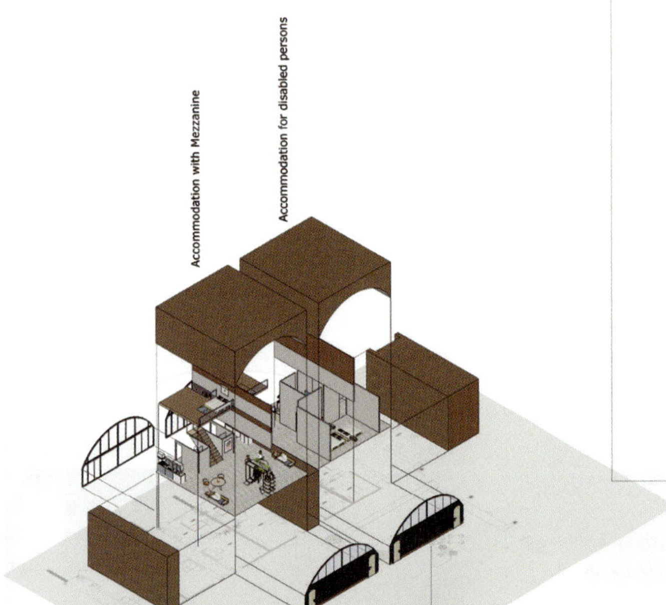

The charred timber

The charred timber specified for the cladding will be FSC certified and the charring process adds significant durability and weather protect to the cladding, which means that it does not need to be treated or painted year on year.

피복재용으로 지정된 숯재는 FSC 인증을 받고, 충전 공정은 피복재에 상당한 내구성과 날씨 보호를 더하여 매년 가공 또는 도장할 필요가 없다.

Reclaimed Antique white oak

Sea grass carpet

Open plan living area for disabled persons on single level. The kitchen is designed to allow wheelchair access and is positioned at a lower level.

My design was inspired by Japanese philosophy Wabi-Sabi : The aesthetic is sometimes described as one of beauty that is "imperfect impermanent, and incomplete"

Using natural clay plaster for the decorative wall finish which has a neutral colour, the texture is not shiny and has a roughness to it.

Using light circular warm yellow clay plaster as a representation of the warmth and sunshine.

The floor is made using the recycled oak timber to create and atmosphere of warmth & movement.
13

저층에 거주하는 장애인을 위한 열린 계획 생활 공간. 부엌은 휠체어에 접근할 수 있도록 설계되었으며 낮은 층에 위치한다.

나의 디자인은 와비사비(Wabi-Savi)의 영향을 받았다. 미학은 때로 "완벽하고 불완전하다" 는 미의 하나로 묘사된다.

중립적인 색상의 장식 벽 마감에 천연 점토 회반죽을 사용하는 것은 광택이 나지 않고 거칠기 때문입니다."

밝은 원형으로 된 따뜻한 노란색 점토 회반죽의 사용은 따뜻함과 햇빛의 표현

바닥은 따뜻함과 움직임의 분위기를 만들 위해서 재활용된 오크목재를 이용하였다.

William Turner
Interior Architecture and Design
Studio 2 NEXUS / Year 3

Pedley Street
Pedley Street is in an area which has been isolated from the busy East End districts such as Brick Lane and Shoreditch from the entensive railway developments at the beginning of the 20th ccentury and again in 2016 when Newtwork rail extended the overground
across London.

페들리 스트리트는 광범위한 철도개발이 이루어진 20세기 초 브릭레인과 쇼어디치 같은 혼잡한 이스트 엔드로부터 격리된 지역이었고 그리고 다시한번 2016년에도 뉴트워크 레일이 런던을 가로질러 가는 지상으로 확장했다.

Briefing challenge:
The current location feels overwhelmingly unloved. The project faces the challenge to turn the space into somewhere that feels welcoming and homely

현재 위치는 압도적으로 사랑받지 못한다는 느낌이 든다. 이 프로젝트는 공간을 전환하여 사람들을 환영하고 집안과 같은 분위기를 전달할 수 있는 그러한 장소로 바꾸는 도전에 직면하고 있다.

The Site is situated within both depot and rail arches owned by Network Rail which are now unused and either boarded up or used as lock uo spaces for old cars

현재 그 장소는 네트워크 레일 소유의 창고와 레일 아치 양쪽에 위치하고 있고 사용하지 않고 판자로 막혀 있거나 오래된 자동차의 보관공간으로 사용된다.

The Highline running alongside the railwayline has an unique and exciting energy

철도를 따라 흐르는 하이라인은 독특하고 흥미로운 에너지를 가지고 있다.

Risen above ground level, the space has a rare sense of escape in the heart of the city, with astonishingly rare views of the skyline.

지상 위로 솟아 오른 하이라인은 놀라울 정도로 보기 드문 스카이라인과 함께 도시의 심장부에 흔치 않은 탈출감을 가지고 있다.

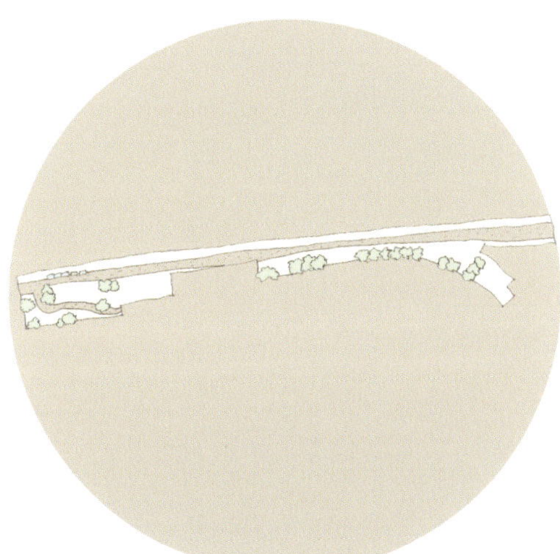

Pedley Street I

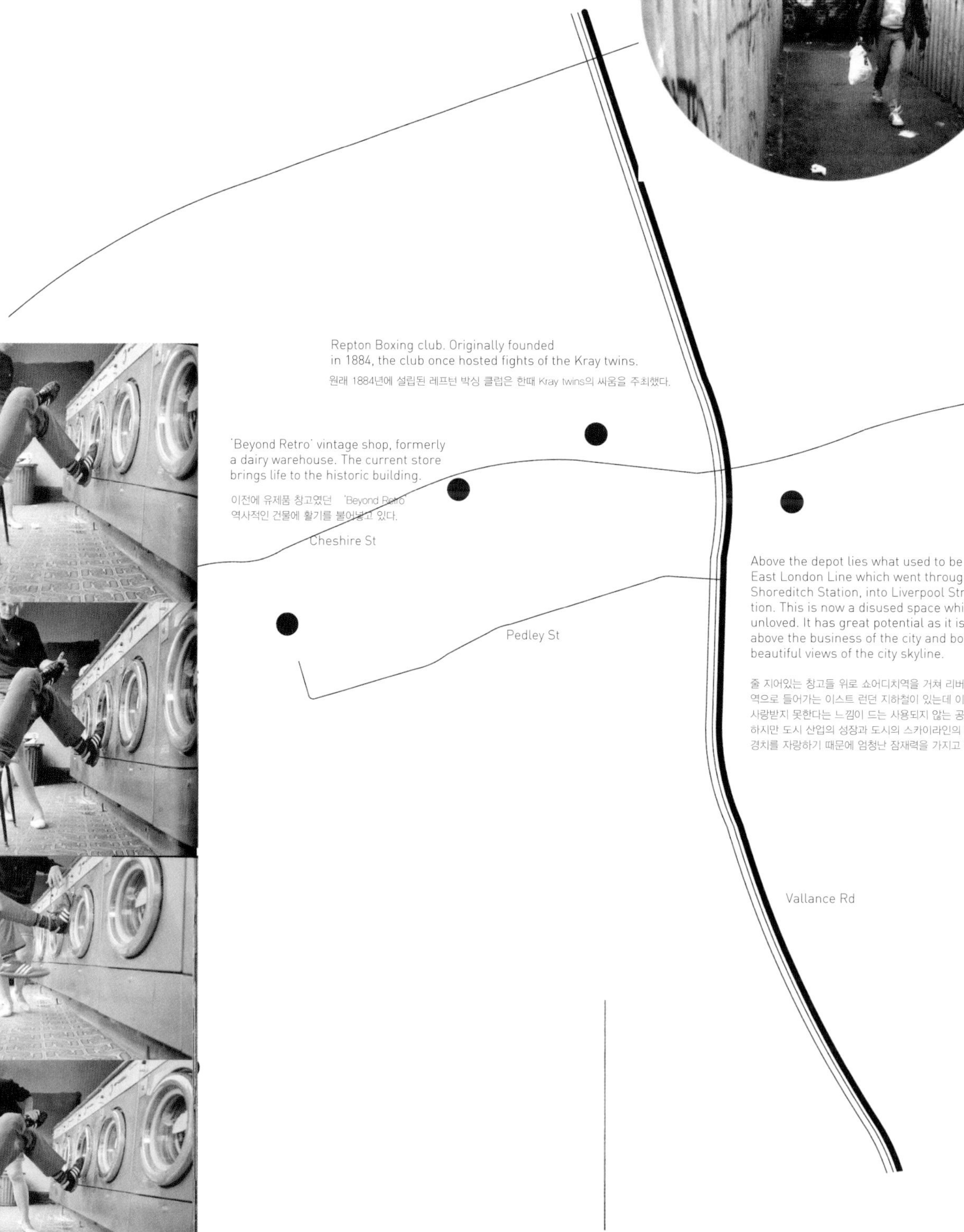

Repton Boxing club. Originally founded in 1884, the club once hosted fights of the Kray twins.

원래 1884년에 설립된 레프턴 박싱 클럽은 한때 Kray twins의 싸움을 주최했다.

'Beyond Retro' vintage shop, formerly a dairy warehouse. The current store brings life to the historic building.

이전에 유제품 창고였던 'Beyond Retro' 역사적인 건물에 활기를 불어넣고 있다.

Above the depot lies what used to be the East London Line which went through Shoreditch Station, into Liverpool Street Station. This is now a disused space which feels unloved. It has great potential as it is risen above the business of the city and boasts beautiful views of the city skyline.

줄 지어있는 창고들 위로 쇼어디치역을 거쳐 리버풀 스트리트 역으로 들어가는 이스트 런던 지하철이 있는데 이곳은 이제 사랑받지 못한다는 느낌이 드는 사용되지 않는 공간이다. 하지만 도시 산업의 성장과 도시의 스카이라인의 아름다운 경치를 자랑하기 때문에 엄청난 잠재력을 가지고 있다.

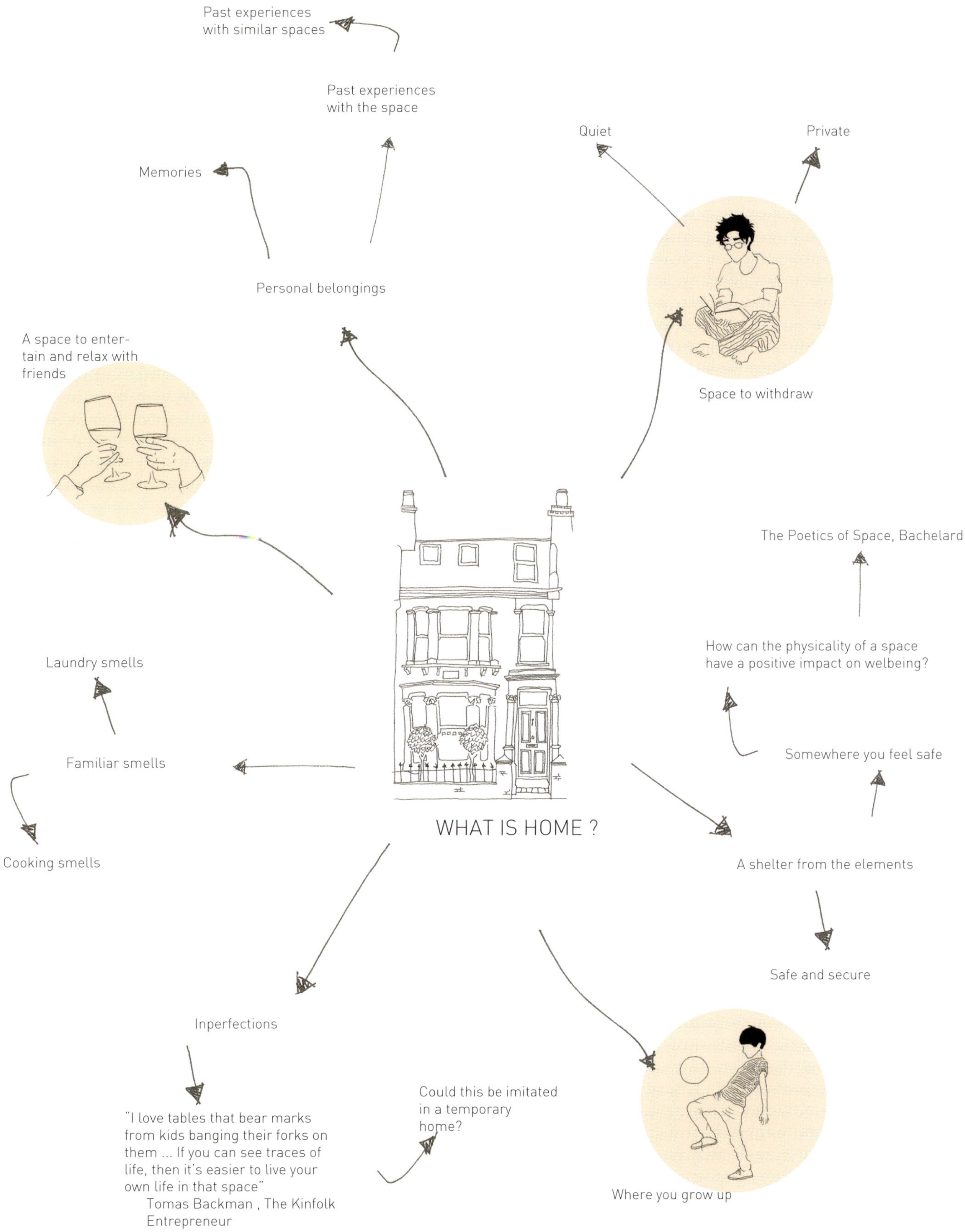
WHAT IS HOME ?
Personal belongings
Memories
Past experiences with the space
Past experiences with similar spaces
Space to withdraw
Quiet
Private
A space to entertain and relax with friends
The Poetics of Space, Bachelard
How can the physicality of a space have a positive impact on welbeing?
Somewhere you feel safe
A shelter from the elements
Safe and secure
Laundry smells
Familiar smells
Cooking smells
Inperfections
"I love tables that bear marks from kids banging their forks on them ... If you can see traces of life, then it's easier to live your own life in that space"
Tomas Backman , The Kinfolk Entrepreneur
Could this be imitated in a temporary home?
Where you grow up

Modelling & development

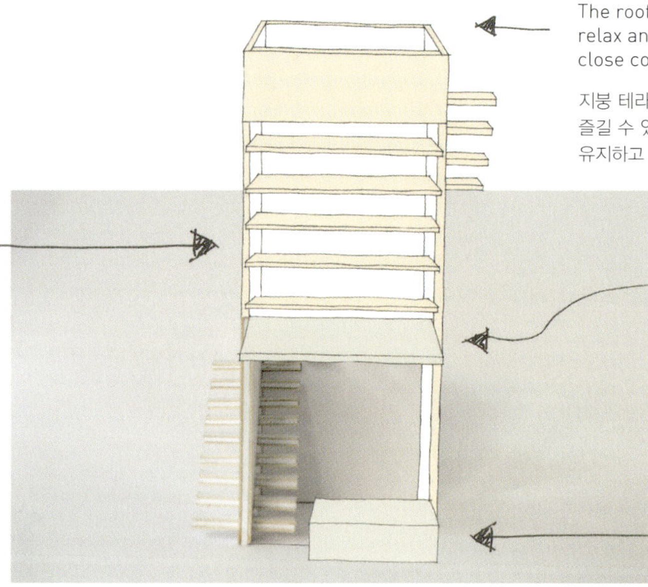

The roof terrace is a space for the occupant to relax and reflect in the summer, still keeping a close connection to the city.

지붕 테라스는 여름에 거주자들이 긴장을 풀고 햇빛을 즐길 수 있는 공간으로, 여전히 도시와 긴밀한 관계를 유지하고 있다.

The front wall of the first floor will be composed of rotating horizontal slats, creating a versatile space.
This offers the occupant a private space that is still connected to the outside; but also has the capacity to become a private and secluded space when needed.

1층 앞벽은 다용도 공간을 만들 수 있는 회전하는 수평슬래트로 구성되어 있다.
이는 거주자들에게 외부와 여전히 연결되어 있는 개인 공간을 제공하지만, 필요할 때 비공개 및 격리된 공간이 될 수도 있습니다.

The upper half of the front wall will fold up to form a canopy over the bench, providin shelter from sun and rain.

앞벽의 윗부분은 태양과 비로부터 보호막을 마련할 수 있는 접혀질 수 있는 벤치 위에 캐노피를 형성할 것이다.

The lower half of the front wall will fold down into a bench, creating an open and welcoming space at ground level.

앞쪽 벽의 아래쪽 절반은 벤치 안으로 접혀 들어가 지상 층에서 열린 환영 공간을 만들 수 있다.

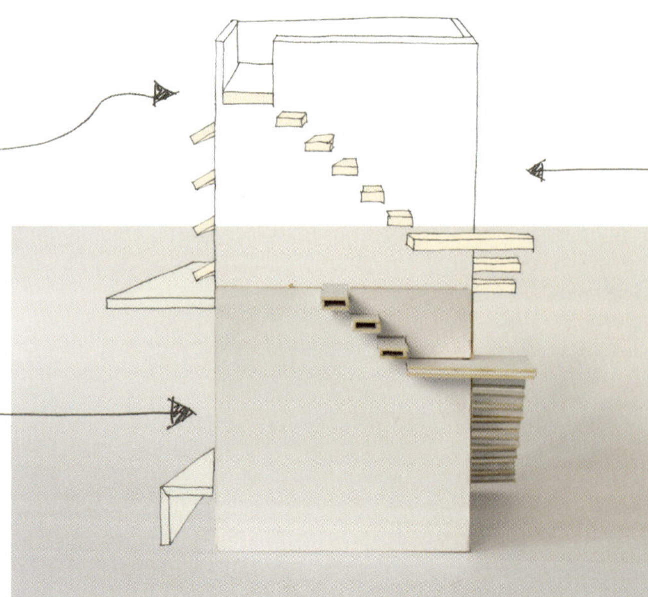

Having the steps around the outside will save space on the inside.

바깥쪽에 계단이 만들어서 안쪽의 공간을 절약할 수 있다.

The first floor will be a livng space. The small space must accomodate cooking, eating, sleeping and relaxing.

생활공간이 될 1층은 작은 공간으로 요리, 식사, 수면 및 휴식을 수용할 수 있어야 한다. 지상층은 거주자에게 제조할 공간과 판매 플랫폼을 제공하는 작업장으로서 기능할 것이다.

The ground floor will function as a workshop, providing the occupant with a space to make and a platform to sell. The space will be able to facilitate a small business, a similar size to a market stall.

그 공간은 시장 가판대와 비슷한 크기의 작은 사업을 촉진할 수 있을 것이다.

Modelling & development

The frame provides privacy to the indoor/outdoor residential space, transforming the forst floor into a secluded garden with views over the city.

이 프레임(틀)은 실내/외 주거 공간에 대한 프라이버시를 제공하며, 1층을 도시 전체의 전망이 있는 외딴 정원으로 변형시킨다.

A frame at the rear of the pod provides the facilities to grow food, allowing the occupant to become self sufficient.

Pod 뒤쪽에 있는 프레임(틀)은 음식을 재배할 수 있는 시설을 제공하여 탑승자가 자급자족할 수 있도록 한다.

The ceiling will be formed of a series of slats which will fold back to create an indoor/outdoor space on the first floor.

천장은 1층에 실내/외 공간을 만들기 위해 뒤로 접히는 일련의 슬래이트로 구성된다.

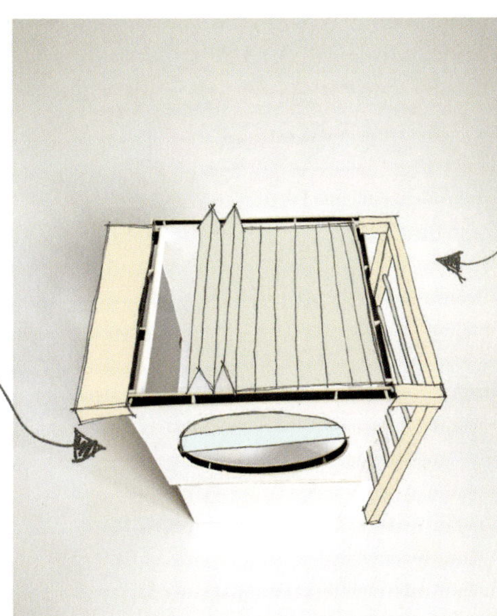

The round window was inspired by the traditional japanese moongates. These are often displayed at the entrances to traditional japanese homes as symbol of welcome.

둥근 창문은 전통적인 일본의 달문들에 의해 영감을 받았고 이것들은 종종 일본 전통 가옥의 입구에 환영의 표시로 전시된다.

Environmental Investigation

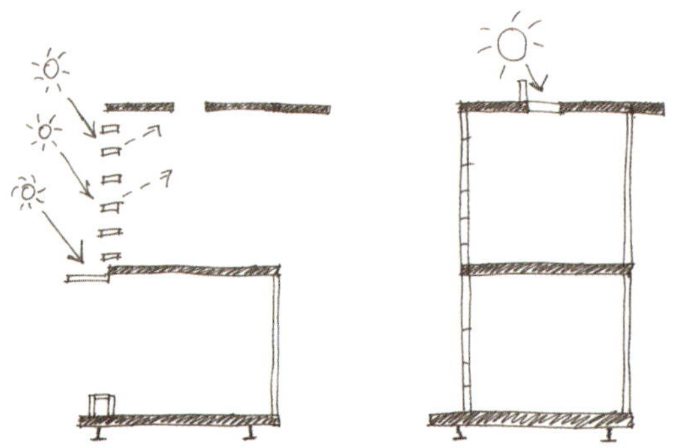

A skylight on the roof of the pod allows light to flood into the space whenm the shutters are closed. The wide shutters provide shade when open and keep the space cool in the summer.

Pod 지붕의 천창은 셔터가 닫힐 때 빛이 공간으로 쏟아질 수 있도록 해야 한다. 넓은 셔터는 열려 있을 때 그늘을 제공하고 여름에는 공간을 시원하게 유지한다.

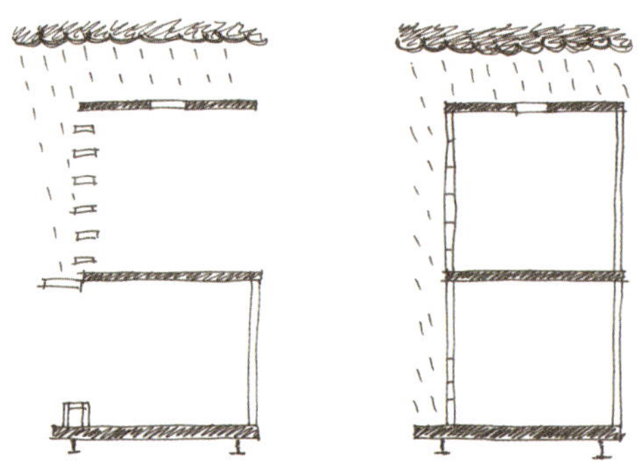

The wide shutters can be angled so that they can remain open during a rain shower in desired. There should be an overlap on the shutters so ensure that the space is completely watertight during heavy rain showers.

넓은 셔터는 원하는 대로 소나기가 내리는 동안 열린 상태를 유지할 수 있도록 각을 맞출 수 있어야 한다. 셔터에는 겹치는 부분이 있어서 폭우가 쏟아지는 동안 공간이 완전히 물이 새지 않도록 빈틈이 없어야 한다.

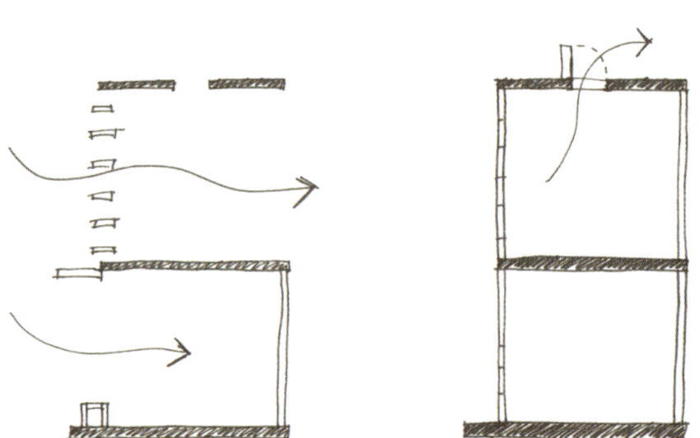

The shutters and the wide sliding door at the rear of the first floor allow breeze to pass through the space, makimng it feel open and less claustrophobic. In addition, the skyight will be open allowing for additional ventilation,

1층 뒤에 있는 셔터와 넓은 슬라이딩 문은 바람이 공간을 지나갈 수 있게 해 주므로 시원하게 느끼고 밀실 공포증을 덜 느끼게 한다. 더하여서, 추가적인 환기를 위해 스카이라이트가 열릴 것이다.

Design Decision

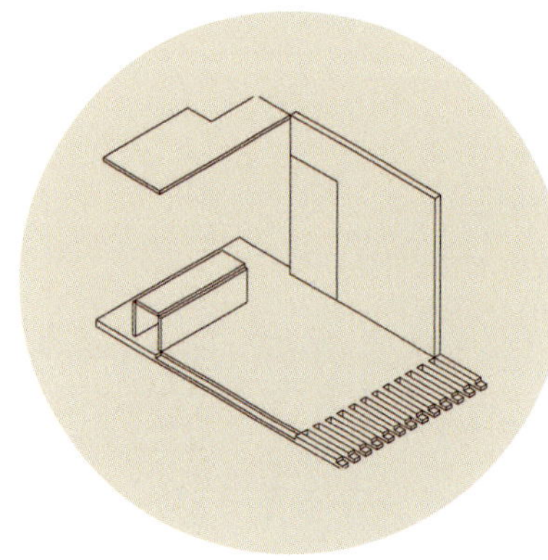

To give the front door enough height, a small section of the canopy must be lost.

프론트 도어의 높이를 충분히 높이려면 캐노피의 작은 부분을 희생해야 한다.

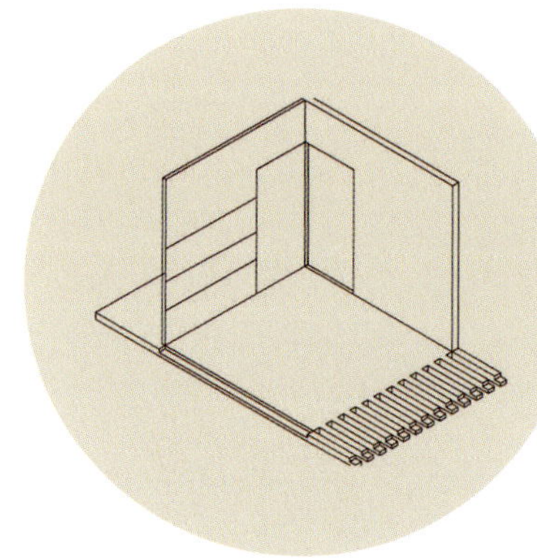

This will compromise no only the size of the canopy, but the look and symetry of the pod design.

이것은 캐노피의 크기뿐만 아니라 포드 디자인의 모양과 대칭도 손상시킨다.

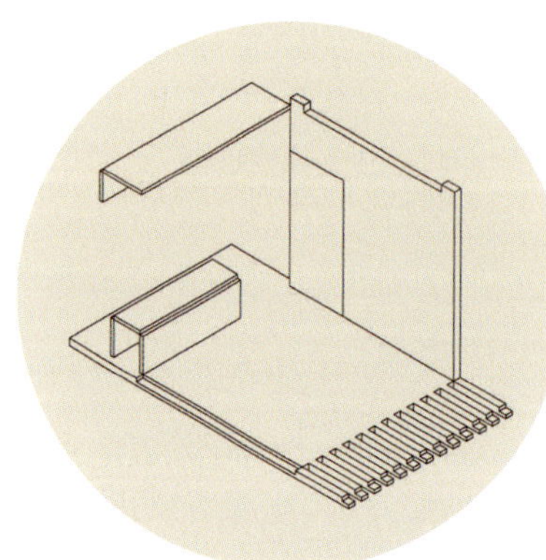

Alternatively, the whole canopy could be cut to fit the door height. The remaining material could be incorporated into a sign for the buiness which will hang from the edge of the canopy when the structure is open.

그와 반해서 전체 캐노피는 도어 높이에 맞게 절단되어야 한다. 나머지 자재는 구조물이 열려 있을 때 캐노피 가장자리에 걸려 있는 사업용 간판에 통합될 수 있다.

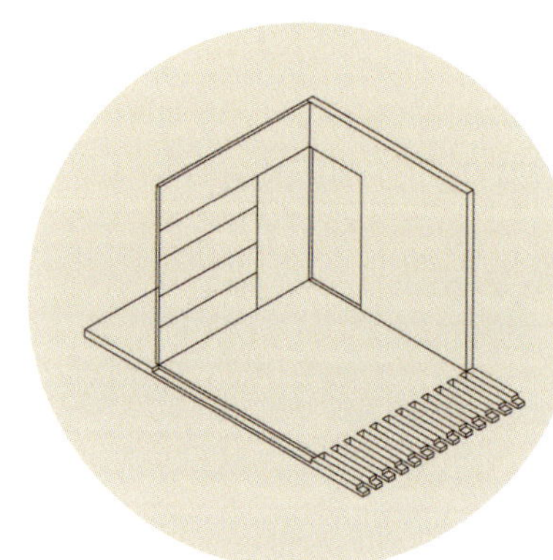

When the structure is shut, the sign would still be visible and positioned in the centre of the front wall.
However, this proposal compromises the depth of the canopy.

구조물이 닫힌 후에도 표지는 계속 보이면서 전면 벽의 중앙에 위치한다. 그러나 이 제안은 캐노피의 깊이를 손상시킨다.

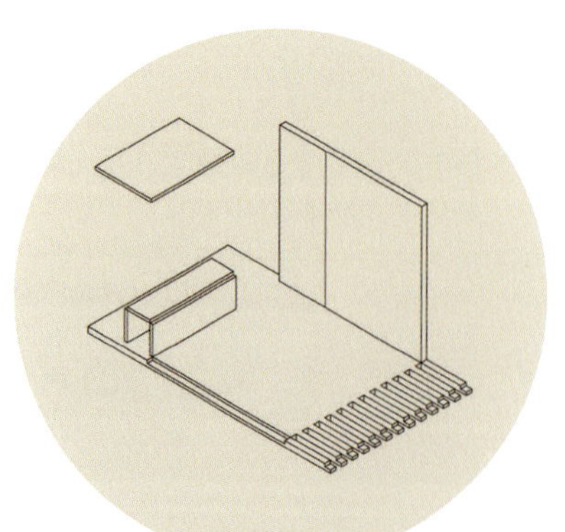

The final proposal is to extend the door height to the ceiling, leaving a canopy over the bench only.

마지막 제안은 천장까지 문 높이를 확장하는 것이며, 캐노피는 벤치 위에만 두는 것이다.

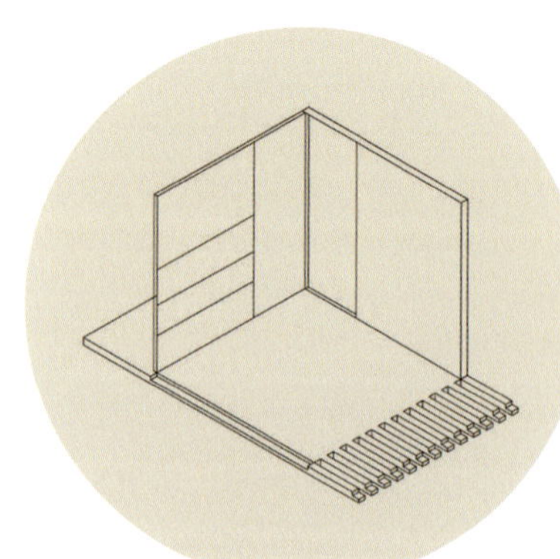

This solution also leaves an aesthetic geometric form when the structure is closed.

이 방법은 구조물이 닫힐 때 미학적 기하학적 형태를 남긴다.

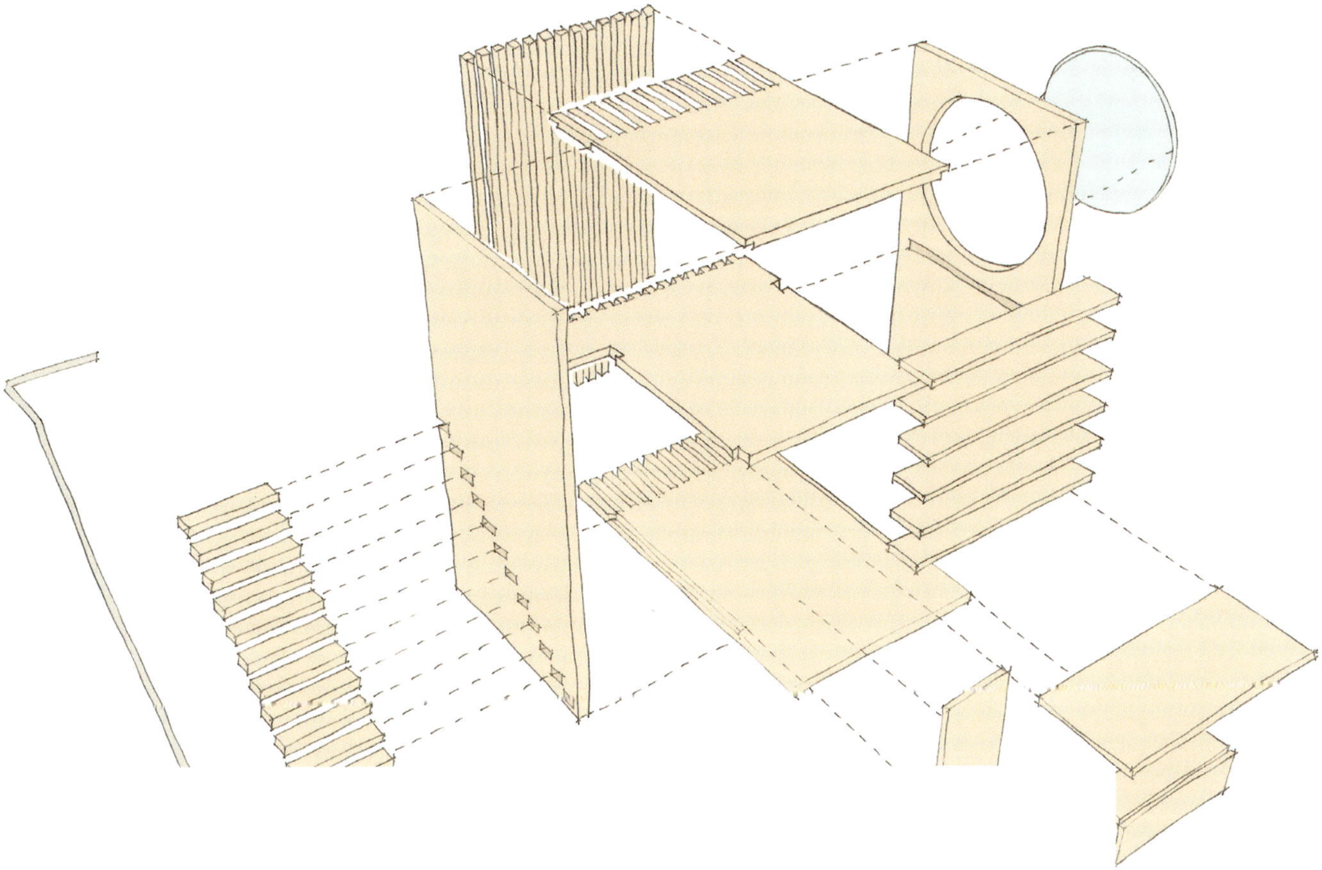

The pod will be constructed from cross laminated timber. The structure of this material is similar to plywood and so will allow for the walls to be manufactured as a single component. This will make the manufacture, distripution and assembly of the pods simple and efficient.

Pod는 교차 적층 목재로부터 만들어질 것이다. 이 재료의 구조는 합판과 유사하므로 벽이 단일 구성 요소로 제조될 수 있는데, 이렇게 하면 Pod의 제조, 유통 및 조립이 간단하고 효율적으로 이루어진다.

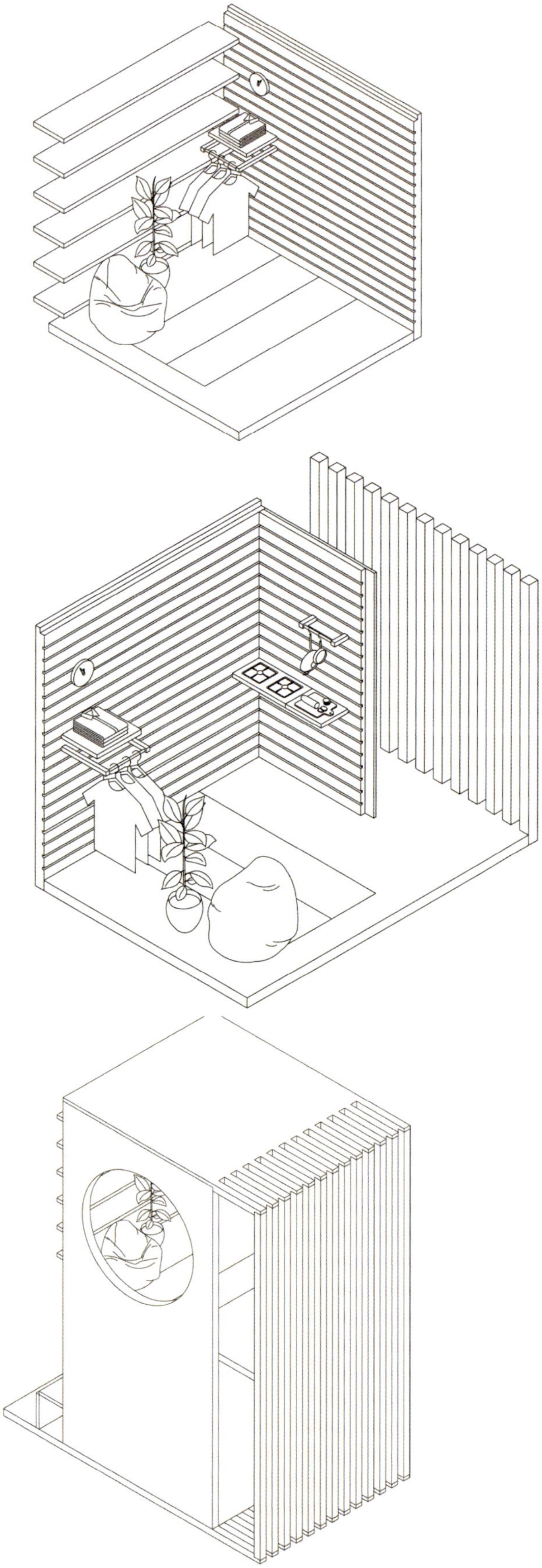

Plan

Rear Isom First Floor Interior

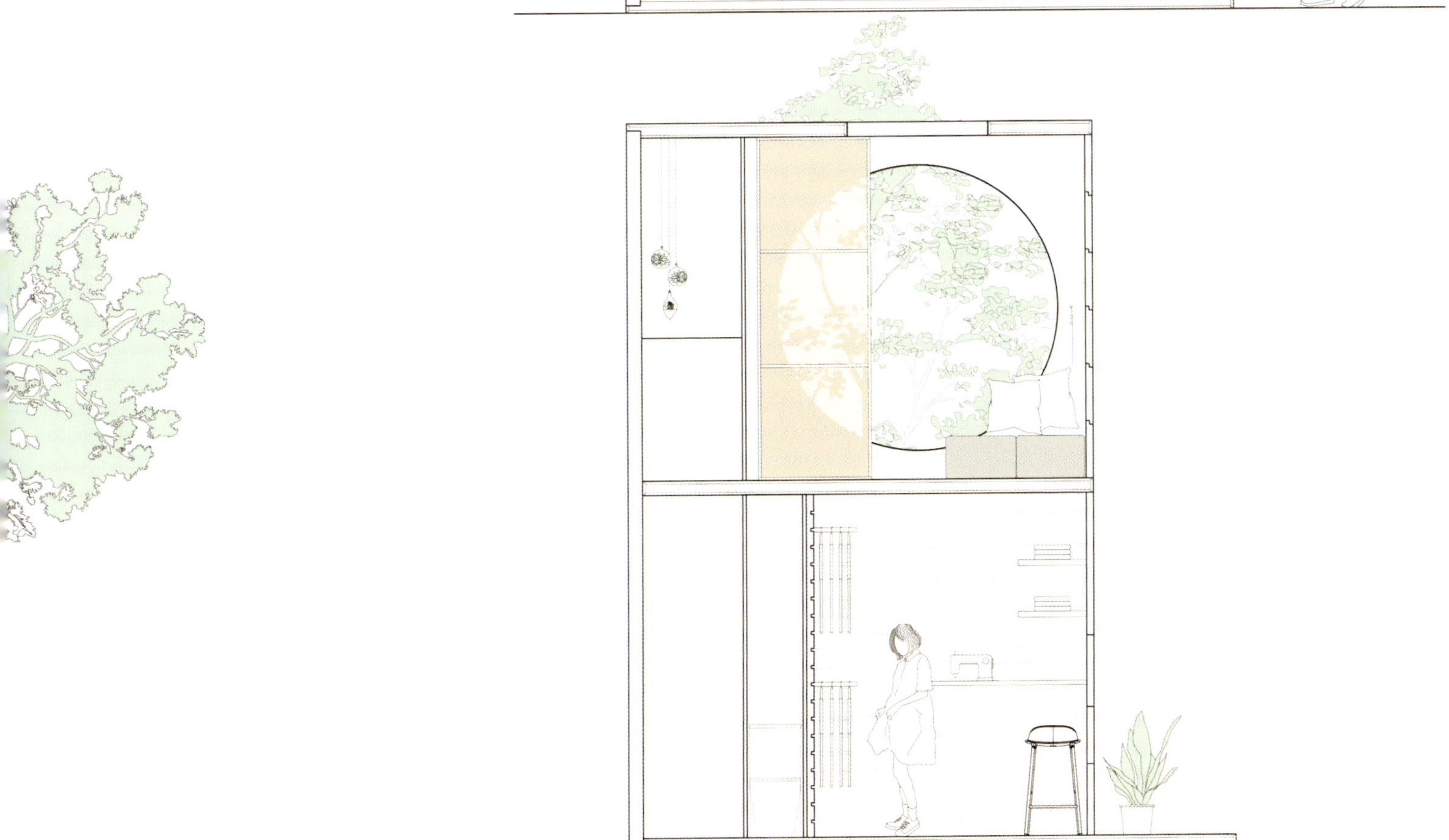

2

DONGYANG MIRAE University
interior & environmental design

DONGYANG MIRAE
UNIVERSITY

DONGYANG MIRAE UNIVERSITY

Department of Interior Environmental Design

2017 sophomore design studio

The 9th Hidden Space Project

RECUSED; Space & People

동양 미래대학교 실내환경디자인과
2017 2학년 2학기 설계스튜디오 결과물

152-714
DONGYANG MIRAE University, Dept.of Interior Environmental Design
62-160, gocheok-dong, guro-gu, seoul, korea
phone + 82 2 2610 1962
fax + 82 2 2610 1960

http://interior.dongyang.ac.kr

Pétur Thomsen, <Imported Landscape> AL3_9a

2학년 설계담당 **박영태 교수**
동양미래대학교 생활환경공학부 실내환경디자인과

responsible for sophomores | Prof. Park, Young-Tae
Department of Interior Environmental Design, DONGYANG MIRAE Univesity
robotomy@naver.com

히든 스페이스 프로젝트 아홉 번째 에피소드
RECUSED; Space & People

자연으로부터, 그리고 자연을 대하는 인간으로부터, 그리고 그러한 인간들 사이의 이념과 자기유지의 이기적 만행으로부터 이미 전지구적인 재난의 시대다. 어쩌면 해 묵은 것이기도 하지만, 결코 해결될 수 없기에 늘 염두에 두어야만 하는 주제다.
플리츠커 상을 수상한 토요 이토, 반 시게루, 그리고 최근에는 쿠마 겐코를 비롯한 많은 건축가들에 이르기까지 자연재해 경험이 많은 일본에서 보여주는 태도와 물리적 구축의 해법은 관련 전공자들에게 좋은 사례가 되고 있다. 세월호 사태이후 각종 자연재해와 인공물에 대한 심각성이 커지면서 2년 전에는 정림건축 문화재단에서 '재난건축' 을 주제로 학생공모전을 진행하기도 하였다.

이렇게 관심이 적지 않은 테마라는 것은 어쩌면 최근 가속화된 자연환경의 위험만큼이나, 그것들을 대하는 우리의 '인식' 에 대한 고민이 많아서라고 생각된다. 인간 문명내부 '실수' 에 의한 재난들, 동시에 대규모 테러나 종교와 이념의 갈등, 독재정권이나 경제적 붕괴로부터 발생하는 수많은 난민(수용)에 대해 우리들이 보여주는 다양한 태도들도 생각해본다. 어쩌면 재난은 인간 역시 자연의 일부라 모두 자연의 피해라고 할 수 있기에, 즉 이길 수 없는 자연이 우리에게 감행(보답)하는, 인간의 감당여부와는 무관한 냉혹한 선물인 것 같다.

이번 설계 스튜디오 수업진행에 있어서 고민되었던 부분은 구체적인 재난상황의 규정과 조건 제시를 통한 문제해결에 집중할 것인지, 아니면, 문제해결 자체보다는 폭넓은 인식적 전환으로서, 즉 체험과 전시연출을 통한 '재난' 에 대한 일상에서의 예민한 감각 회복에 집중할 것인지였다. 문제해결은 재난 발생이라는 실제 상황이라는 예측 불가한 변수에 대응하는 것이기에 기술적인 한계가 자명한 것이었고, 또 재난 이후 진행될 사후적 처치로서 진행되는 물리적 공간의 대안 역시 실제 일본건축(가)가 보여준 방법이나, IKEA사에 보여준 조립식 건물의 사례 등을 보았을 때 이번 수업을 통해 그 이상으로 획득될 생산적 여지가 여간해서 떠오르질 않았다.

Hidden Space project Episode 9
RECUSED; Space & People

It is already an age of global disaster from nature and from the human beings who treat nature and from the selfish barbarism of ideology and self-sustaining among such humans. Maybe it's something we've done so far, but it's something we should always keep in mind because it just can't be solved.

The attitudes and physical construction solutions shown in Japan, ranging from Toyo Ito and Ban Sigeru who took the pritzke prize to many architects recently Kuma Genko, are a good example for related majors. After Sewol ferry disaster in 2014, as the seriousness of various natural disasters and artifacts grew, the Chungrim Cultural Foundation held a student exhibition two years ago under the theme of "Disaster Architecture."

TThe topic is probably just as much concerned about our 'awareness' about theme as it is about the dangers of recently accelerated natural environments. Also, it is considered the various attitudes we show about disasters by 'mistake' within human civilization, and the numerous displaced people (refugees) resulting from massive terrorism, religious and ideological conflicts, dictatorship or economic collapse. Disaster is a kind of cruel gift that is not only human can't beat but also whether human can overcome because it is a part of nature.

[illegible] of professor, Won sungwon, C-Print [illegible] cm 2017

각 팀별로 재난의 개념을 정의 -> 공간디자인 수행의 측면에서 재정의.

1. 재난발생당시 인간의 안전을 담보할 수 있는 공간, 혹은 사물(장치)의 기능적 수행력–문제해결

2. 재난발생 이후 생활유지 공간의 기능적 수행– 문제해결

3. 재난에 대한 전시, 체험 기능–문제에 대한 교육, 예방과 치유의 공간

따라서 팀별 상황과 조건은 최대한 자유롭게 설정하였다. 재난의 정의설정으로부터 사회문화적 흐름 안에서 각 팀별 관심사를 도출하였다. 물리적 재난만큼이나, 사회적 재난의 관심이 많았다. 현대성(인간소외)에 본질적 질문을 던지는 것들이 많았는데, 이미 충분히 소외된 재난의 강도와 질적 차이를 염두에 두며, 그것들이 최대한 공간화 될 수 있도록 논의하였다. 재난안전관리 기술에 대한 전문가의 특강을 통해 "재난" 의 실체와 문제해결방식, 그리고 그것들을 전시하는 방법으로 가상현실, 증강현실의 사례를 학습하였다. 소방서, 경찰서, 파출소, 응급센터 등 재난업무를 수행하는 행동가들을 방문 인터뷰하였으며, 팀별 재난상황에 내재된 특수성을 최대한 낱낱이 해부하였다.

What was the main point about this project was whether to focus on solving problems through specific disaster situations and condition planning, or as a broader cognitive transition which is about the recovery of a sensitive sense of 'disaster' through experience and exhibition rather than problem solving. Especially, in terms of problem solving, it is hard to acquire productive solutions through this project because of unpredictable variable of disaster occurrence even though there are two example solutions which are the alternative of physical space as a follow-up treatment demonstrated by the actual Japanese architecture and a prefabricated building presented from the IKEA Company.

Define the concept of disaster for each team --> Override in terms of space design performance

1. Space to guarantee human safety in the event of a disaster, or functional performance of objects (devices) -> problem solving

2. Functional performance of the living space after the disaster -> problem solving

3. Exhibition and experience of disasters -> the space of education, prevention and healing on a problem

Therefore, the conditions and situations of each team were set as freely as possible. From the definition of a disaster, the concerns of each team were derived within the sociocultural flow. There was as much interest in social disasters as in natural disasters. There were many fundamental questions to modernity (the alienation of human being), and the intensity and qualitative differences of disasters that were already excluded were discussed to maximize their spatiality. Special lectures by experts on disaster safety management technology taught the real world of "disaster" and problem-solving methods and examples of virtual reality and augmented reality by displaying them. We visited fire stations, police stations, police offices, emergency centers, and interviewed the workers who carried out the disaster and dissected the specialities inherent in the disaster situation as much as possible.

총 11개의 작품 진행을 위해 마무리까지 노력한 학생들 개개인에 고마움을 전한다. 어려운 (기술적) 문제해결만큼이나, 재난이라는 상황을 너무 손쉽게 타자화하지 않아야 하는, 따라서 상상력을 충분히 발휘하기 힘든 프로젝트였기에 아쉬움이 남는다. 이번에도 온라인으로만 교감할 수밖에 없었지만, 항상 새로운 관점을 모색하는 데에 열정을 보이는 런던 메트로폴리탄 대학의 Kaye Newman 교수와 학생들에게도 감사드린다.

I really appreciate each student for trying to finish 11 works. It is regrettable that it was a project that was difficult to fully use imagination as we cannot easily imagine the situation of disaster. I had to communicate online again this time, but I am also grateful to Professor Kaye Newman and students at the University of London who are always passionate about finding new perspectives.

Module Summary (2017.09.06)

01 REQUIRED COMPULSORY SUBJECTS

Sophomore "Inteiror Design Studio III"

4 Of 18 Credit Modules in a Semester
- 6 Hours (tuesday) / 1 Week
- 1, or 2 Students / 1 Team
- Total 11 Projects

02 THEME

"REINTERPRETATION"

03 PROCESS & OUTPUT
- Research & Analysis (include FGI & survey, SWOT)
- Space Scenario making
- Sketch(analog & digital method) process
- Complex Space Program(zoning, planning....)
- Drawing
- 3D Simulation
- Scale Modeling
- Panel Presentation
- & Every Process(Photo & Text) Recording
  - for the Final Documentation

04 SCHEDULE
- 2017.9. 1st Week - Kick off & Orientation
- 2017.10.Last Week - Mid Term Critic Exhibition & Panel presentation
- 2017. 12. Last Week - Final Exhibition

05 KEY WORD
- DISASTER
- NATURAL DISASTER
- DISTRESS
- RESCUE
- NATURE & CIVILIZATION
- REFUGEE
- DYSTOPIA
- MATERIALS & ENERGY
- CULTURE CAPITAL
- ART, LABOR, EDUCATION, EVERYDAYNESS

cial lecture 01 - disaster management plan

special lecture 02 - container architecture

Choi, Doo-soo (Artist/Curator in MULLAE) / Lee, Chang-Han(ee-movement, CEO)

1st day_06, september ORIENTATION | The 7th Bldg. 5F AM 09:00~

Kim, Suk-Young(Professor, Kun-kuk UNIV.)

박희영 I Park Hee Young 오지현 I Oh Ji Hyeon

TWELVE & GONG ZAK SO

재난에 대비하는 장애인과 지역주민들을 위한 주거 복합 공간

A residential complex for disabled people and local residents in preparation for disaster

BACKGROUND

후쿠시마 원전사고

후쿠시마 현에 위치한 행정구역으로 원전지대로부터 20km 떨어져 있는 곳에 위치해 있으며 바다와 아주 근접한 땅으로 마치 작은 산을 연상시킨다. 주변에는 기념공원이 있고 언덕으로 올라가면 신사와 불교 사찰을 볼 수 있다. 동쪽으로는 바다가 근접해 있으며 동, 서, 남쪽은 밭이 있는 해안 농경지역이다. 재난이 일어난 이후 현재로는 아무 것도 세워지지 않은 빈 공터이며 산과 연결 되어 있기 때문에 약간의 경사진 대지를 이루고 있다.

Fukushima nuclear accident

On March 11, 2011, an earthquake and a tsunami that penetrated the northeastern part of Japan and caused a plant to leak radioactivity in Fukushima Prefecture. The residents in the affected area have evacuated from Fukushima Prefecture, but there are still people who failed to evacuate and are living there. People who could not evacuate. They are the disabled who is going to die before the disaster. I would like to build a place of communication and connection where the native people can stay together to overcome the disaster that may come again and comfort their wounded hearts

SITE

후쿠시마 현에 위치한 미나미소마시 해안가 언덕
면적 : 2000㎡ (605평)

Minamisoma Coast Hills, Fukushima Prefecture

Area : 2000 m^2 (605 m^2)

SITE ANALYSIS

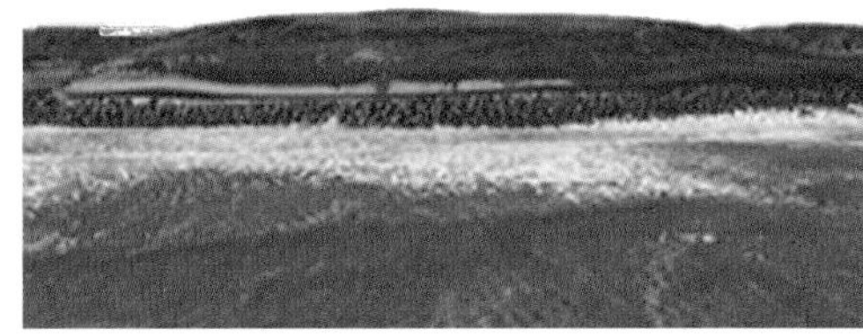

후쿠시마 현에 위치한 행정구역으로 원전지대로부터 20km 떨어져 있는 곳에 위치해 있으며 바다와 아주 근접한 땅으로 마치 작은 산을 연상시킨다. 주변에는 기념공원이 있고 언덕으로 올라가면 신사와 불교 사찰을 볼 수 있다. 동쪽으로는 바다가 근접해 있으며 동, 서, 남쪽은 밭이 있는 해안 농경지역이다. 재난이 일어난 이후 현재로는 아무 것도 세워지지 않은 빈 공터이며 산과 연결 되어 있기 때문에 약간의 경사진 대지를 이루고 있다.

This hill is the administrative district of Fukushima Prefecture, 20km from the nuclear power plant, very close to the sea and it reminds of a small mountain. There is a memorial park nearby and you can see shrines and Buddhist temples on the hill. On the east, there is a sea nearby and the east, west, and the south area are the coast agricultural field. Since the disaster, the land is still empty and is connected to the mountains, having slightly inclined surface.

TARGET

계속되는 분쟁과 폭력으로 인해 교육의 권리를 박탈당한 아이들에게 책에서의 배움(learn)을 통한 학습의 기회를 제공하고, 그들이 죽을 때 까지 경험해 보지 못하는 신비로운 환경을 공간에 실현시켜 놀이(play)와 체험(experience)을 통해 잠시나마 전쟁의 아픔을 이겨 낼 수 있는 즐거운 치료와 소통의 공간이 되었으면 한다. 더불어 이런 공간이 제공됨으로써 공통된 사연을 가진 그들이 서로 의지하며 교감이 나누는 환경이 조성되기를 바란다.

For the children who were educationally deprived from the continuous trouble and violation, I hope to provide the "Learning" opportunity from the book and by realizing the mysterious environment to the space, through the play and experience, i sincerely hope the place where those things can lead to them to endure the pain of war for a while. And also, by providing 'this' environment, i hope that people who have general stories depend each other and share their feelings

CONCEPT

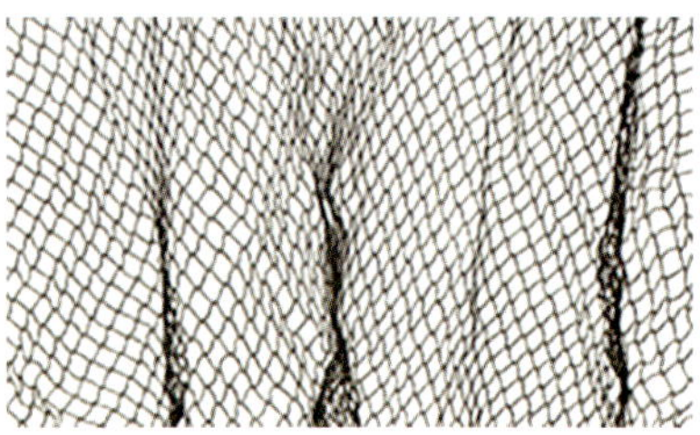

그물 (NET)_NETWORKING SPACE

얇은 실을 재료로 하여 줄과 줄이 서로 얽혀 제 각각인 구멍의 크기를 이루는 그물의 형상.
나누어지지만 서로 얽힌 그물망은 수직적, 그리고 수평적으로 연결되며 짜임을 이룬다. 이러한 그물의 짜임을 바탕으로 형성된 VOID공간 그리고 따로 떨어져 있는 VOID공간을 연결하는 SOLID 통로를 통해서 모든 공간은 하나의 NETWORKING SPACE를 형성한다.

A shape of mesh in which rows are combined with materials to form the size of individual holes. The divided but interlinked mesh consists of vertical and horizontal connections. All spaces form a NETWORKING SPACE through the VOID space formed based on the mesh of this net and the SOLID path connecting the separated VOID space.

장애인들이 스스로 이동하기에 불편함이 없는 1/12 경사로
A convenient 1/12 slope way for disabled people to move by themselves.

휠체어를 타고 타인의 도움 없이 편하게 이동 할 수 있는 경사로 비율 1/12를 SOLID공간에 대입하여 통로로서 공간의 전체를 수평적 수직적으로 연결시킨다. 이로 인해 모든 공간이 막힘없이 연결되고 공간의 흐름에 따라 자연스럽게 시퀀스가 형성된다.

A shape of mesh in which rows are combined with materials to form the size of individual holes. The divided but interlinked mesh consists of vertical and horizontal connections. All spaces form a NETWORKING SPACE through the VOID space formed based on the mesh of this net and the SOLID path connecting the separated VOID space.

DESIGN PROCESS

STEP 01. 1/12 경사에 의한 연결 1/12 slope connection

서로 얽혀진 그물망의 형태와 장애인들이 스스로 이동하기에 불편함이 없는 1/12 경사로를 통해 모든 공간은 막힘 없이 연결된다.

The shape of the interconnected networks and through the 1/12 tracks allow disabled people to move easily, all spaces are smoothly connected.

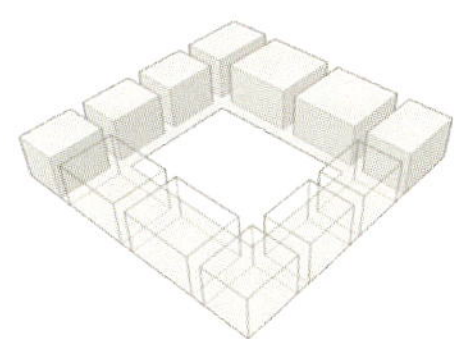

그물의 짜임을 모티브로
분리된 공간 형성
Building a separated space
with the motivation of net system

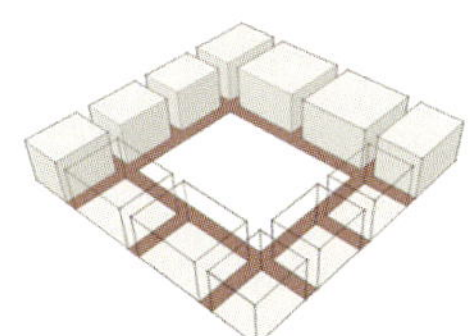

SOILD 통로를 통해
모든 공간을 연결
Connect every space
through the SOILD path

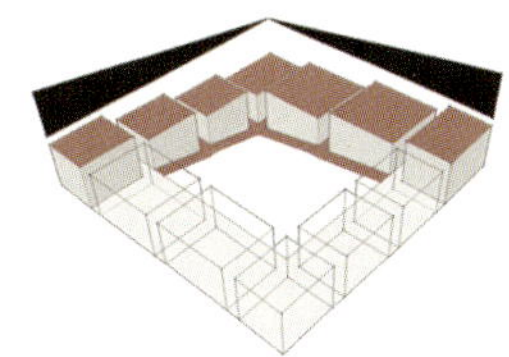

1/12경사로를 대입해
MASS 전체의 높낮이 조절
Adjust the height of the entire mass
by substituting 1/12 slope system

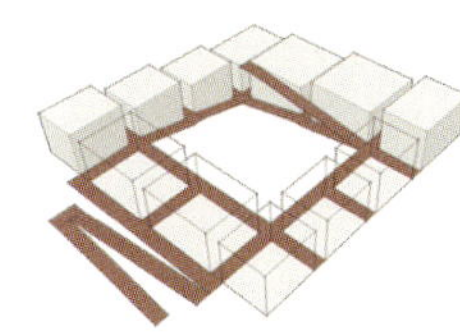

진입로부터 옥상정원까지
모든 공간이 1/12 경사를 따라
수직수평적으로 연결되며 시퀀스형성
Arrangement from the entrance
to the rooftop garden connected vertically
and horizontally along a slope

STEP 02. 공간의 배치 Arrangement of space

장애인들과 그의 가족들이 머무르는 사적인 공간과 공작소, 도서관, 공용주방, 휴게공간으로 형성된 연린공간

Private space and workshop area, library, a public kitchen for the disabled and their families, an open space for rest zone

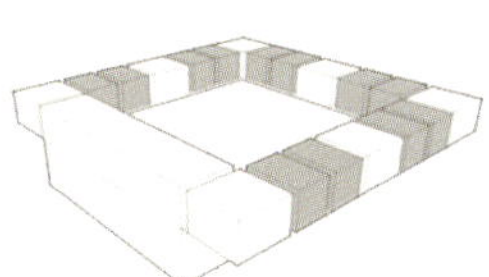

주거 공간 Residential space

12가구를 수용할 수 있는 주거 공간
Residential space that can
accommodate 12 households

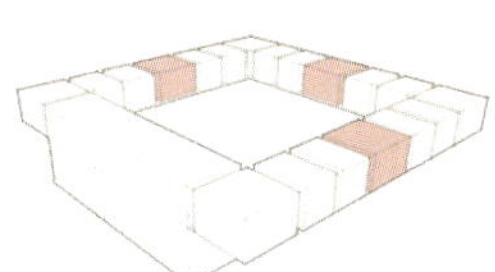

주방 공간 Kitchen space

주거존에 위치하여
사이 사이 배치된 공용 주방 공간
Kitchens located in residential place

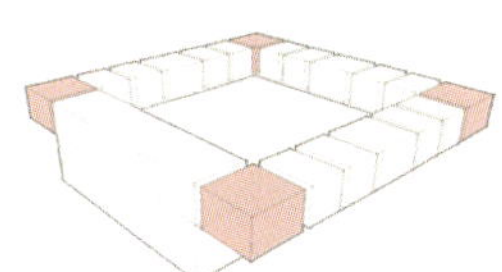

휴게 공간 Rest space

모든 사람에게 열린 휴게공간
Rest area for everyone

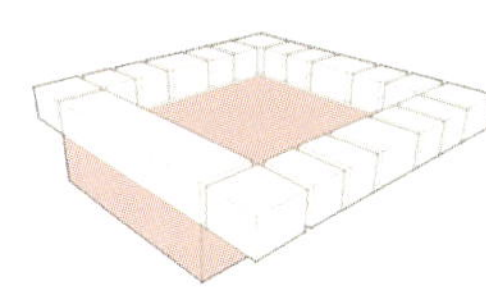

도서관 Library

지하1층에 위치한 도서관 공간
(재난시에 임시대피소로 사용)
The library on the ground
floor of the lounge
(also used as a temporary
shelter in disaster)

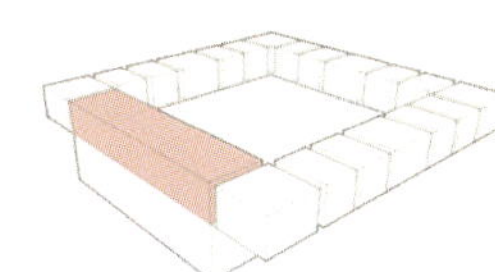

공작소 Gongzakso

주 출입구와 연결되어
모두가 한번씩 거쳐 가는 공작소 공간
Workshop space connected
to main entrance,
which everyone passed at least once

STEP 03. 구조적으로 안정된 기반 Structurally stable foundation

해안지역 특성을 고려하여 바다 바람에 버틸 수 있는 단단한 기반 형성

Considering the characteristics of the coastal area, the structure was formed a solid base that can withstand the wind.

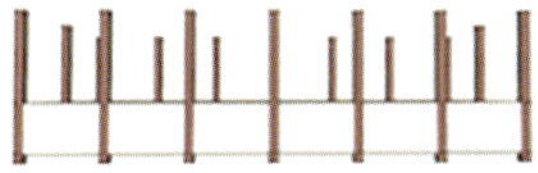

H형강 기둥으로 구조적 틀 마련
Structural framework with H-column

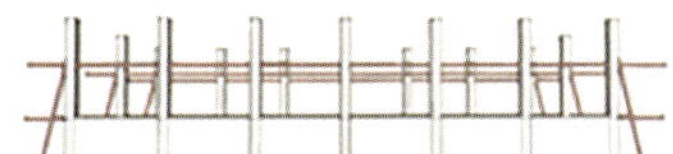

비계결합
Scaffold joint

맞물린 기둥과 지붕틀
interlocked Pillar and roof frame

PROGRAM

공작소 GONK ZAK SO

工作所 : 뚝딱뚝딱 만드는 곳 → 공작소

직접 체험하고 참여함으로써 공작소 내부의 작업실에서 장애인들이 만든 공예 작품을 통로에 전시하고 직접 판매 할 수 있는 공간이다. 장애인들이 스스로 할 수 있다는 경각심을 일깨워 주고 자립을 도와주는 역할을 하며 마을 주민들과 소통 할수 있는 열린 공간이다.

The place where the tuck is made → Gongzakso

Studio- This is the place where visitors can participate and even sell the displayed handicrafts, which the disabled made it in the aisles from the studio. It is an open space for the disabled to realize that they also can do everything on their own, and communicate with the villagers.

空作所 : 비워져 있지만 소통하며 채워질 곳 → 도서관

모두에게 열린 공간으로 거주하는 사람들과 마을 주민들이 함께 채워가는 도서관 공간이다. 얼굴을 마주할 수 있는 개방공간과 사적인 시간을 가질 수 있는 개인공간으로 구성된다. 도서관내의 가구는 1/12의 경사로 비율에서 가져온 사선 축 위에 배치되며 자연스러운 시퀀스를 형성한다.

An empty place but will be filled with communication → Library

Like an open place here is filled with residents. It is composed of two separated places. The first one is an open place and the other one is a private place. Furniture in the library will be placed on a diagonal axis set up at a rate of 1/12 and forms a natural sequence.

公作所 : 함께 공존하는 곳 → 주거존

장애인들과 가족들이 거주 하는 주거공간과 이웃들이 함께 사용 하는 공용 주방공간으로 형성된다. 프라이버시를 존중함과 동시에 주거공간 사이에 배치된 공용 주방공간은 함께 공존하고 소통할 수 있는 기능을 마련한다.

Place where all lives together → Residence zone

This place is formed as a residential space for disabled people and households and a common kitchen space is being used by neighbors as well. The public kitchen is located between residential spaces and makes it easy for people to communicate together.

FLOOR PLAN

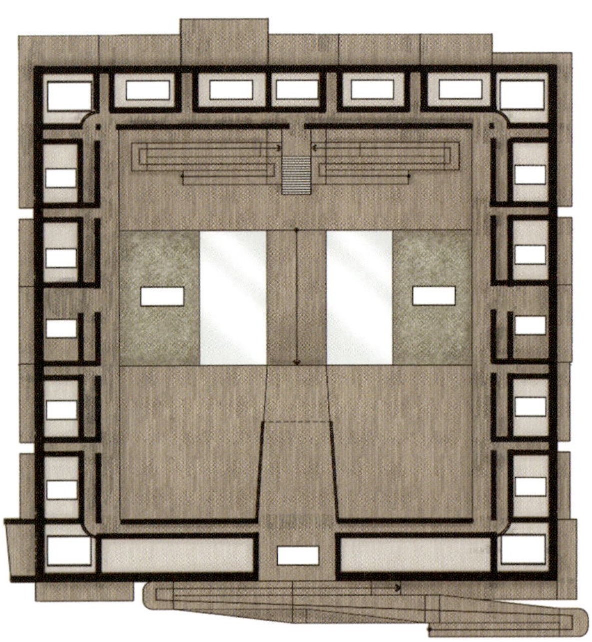

FIRST FLOOR

◀ 공작소를 시작으로 12개의 거주공간과 3개의 공용주방 4개의 휴게시설로 구성된다. 공용주방에는 열린 공간을 형성하기 위해 문을 개방하고 4개의 휴식공간은 건물 사이드에 각각 배치하여 접근성을 원활하게 한다. 공작소는 건물로 진입할때 반드시 거쳐가야하는 공간으로 수평과 수직의 동선이 맞닿아 있다.

With the workshop, it consists of 12 residence spaces and 3 public kitchens and 4 resting facilities. The door is opened to form an open space in the kitchen and four resting spaces are arranged on the side of the building to facilitate accessibility. A workshop is a space that must be passed through when entering this building and has horizontal and vertical movements.

1층에서 경사를 따라 내려오면 도서관과 연결된다. 통로를 중심으로 개방공간과 개인공간 그리고 썬큰공간이 배치되어 있다. 도서관의 전체적인 벽면은 유리창으로 두어 사색속의 시퀀스를 보여준다. ▶

The floor follows the slope and leads you to the library. Open, private, and dark spaces are placed around aisles. we put windows to the entire wall of the library to show a reflective sequence.

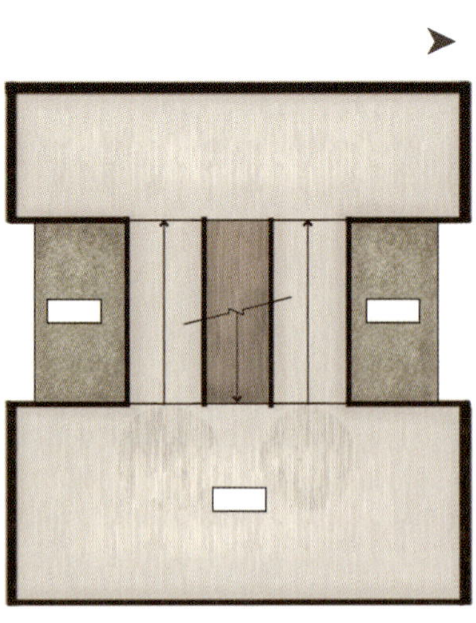

BASEMENT FIRST FLOOR

AXONOMETRIC

1/12경사 통로
1/12 slope passage
도서관 중심에 위치하며
개방공간과 개인공간, 지하도서관과
지상을 연결한다.
It is located at the center of the library and directly connects to the public place, the under-ground library and to the 1st floor.

썬큰가든
Sunken garden
개인공간과 연결되며
1/12 통로 축에 위치한다.
It is connected to the personal space and is located on the the1/12 aisle side.

개인공간
Private place
통로를 중심으로
뒷쪽에 위치한다.
This space is behind the path.

주거공간
Residence area
하나의 주거 유닛에 4가구가
이웃하며 거주한다.
Four households live next door as one dwelling unit.

공용주방공간
Public kitchen
각 주거 유닛에 배치되며
4가구가 함께 사용한다.
Located in each residential place. It will be used by 4 households.

소통공간
Communication space
통로를 중심으로
앞쪽에 위치하며
정면에 뚫린 통창을 통해
외부와 연결된다.
Located in front of the passage and connected to the outside through a window.

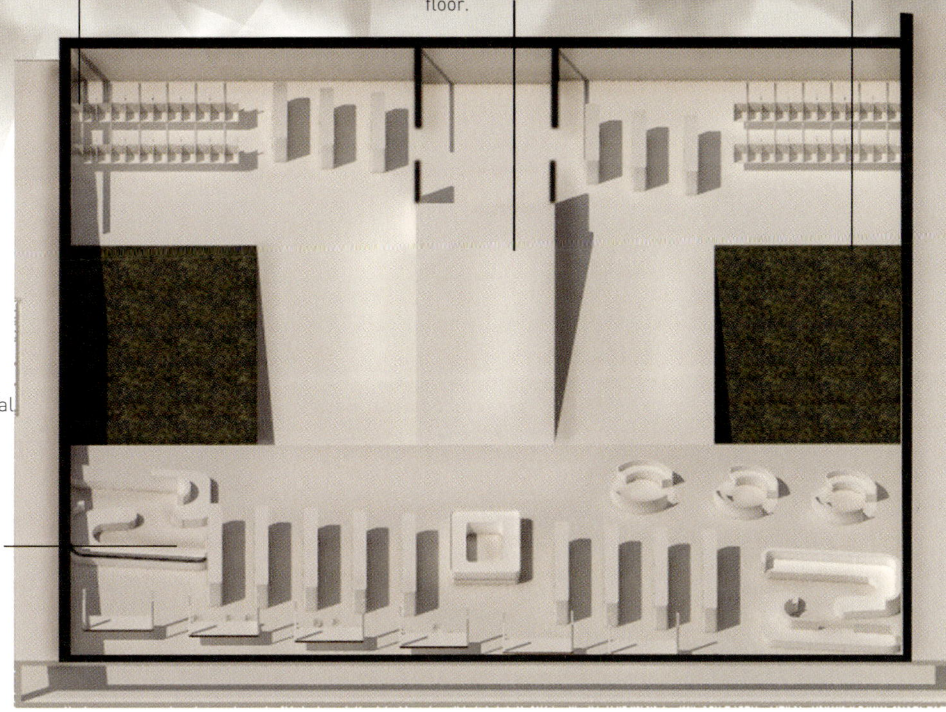

도서관 Library

휴게공간
Resting area
건물의 사이드에 배치되어
어디서든지 접근이 용이하다.
It is located on the side of the building and easy to access from anywhere.

주거존 Residnece zone

전시 통로
Displayed passage
공작소의 모든 공간을 연결한다.
It connects all of the spaces in the workshop.

주출입구이자 판매공간
As the main entrance and sales space
건물의 입구와 맞다은 곳에 위치한다.
It is located in front of the entrance of the building.

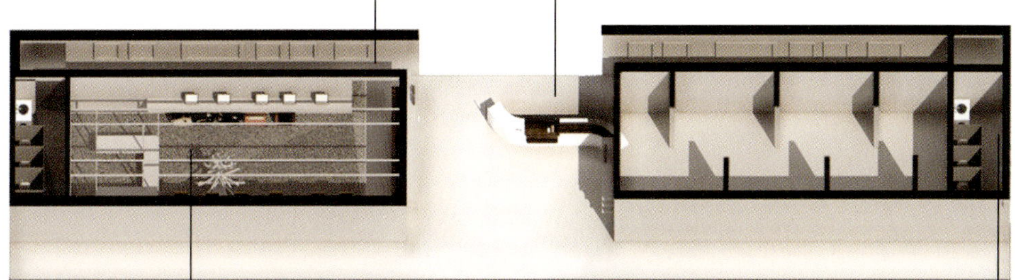

작업실
Workshop
공작소 통로와 연결되어 있다.
It is connected to the workshop passage.

공작소 Gongzakso

공용화장실
Public restrooms
공작소 양쪽 사이드에 하나씩 배치 되어 있다.
Those are placed on each side of the workshop.

BIRD'S EYE VIEW

12 & 공작소는 그물의 짜임을 모티브로 하여 자연스럽게 연결됨으로써 공간이 채워진다. 출입구의 램프로 부터 시작해 전체적으로 1/12의 수평적 경사가 배치되며 자연스럽게 수직적 이동이 가능하게 연결되어 있음을 볼 수 있다. 또 건물은 시시각각 변하는 날씨나 바다바람의 영향을 피하고자 H빔으로 짜여진 구조틀에 지붕 구조를 연결시켰다. 이로 하여금 이곳에 거주하는 장애인들과 시설을 이용하는 사람들은 재난에 대한 불안감을 덜고 시설로부터 보다 안전하게 보호 받을 수 있는 것이다.

12& Gong zak so is filled with the motive of the mesh of the net naturally. Visitors can easily see that starting from the lamp at the doorway, the horizontal slope is entirely located and naturally connected to it. Furthermore, the building was connected to the roof structure which was made with the H beam to avoid the bad influence of the changing weather or the sea wind. As a result of this, residents and the disabled who are living here can be protected

out any anxiety towards disaster.

EXTERIOR RENDERING

건물 진입로인 1/12의 램프를 통해서 지상1층으로 올라가면서 도서관의 실내와 마주하게 되며 도서관에서 또한 정면에 뚫린 창을 통해 외부 진입로가 한 눈에 들어오면서 내부와 외부의 소통을 가능케 한다. 이렇게 1/12 경사에 의해 공간은 연결되며 자연스레 시퀀스가 형성되고 느림의 미학을 느낄 수 있다.

As you walk up one floor of the main building, you would face the interior of the library, and from the library, you can see the way coming in through a window that is right in front of you. Hence, space is connected by the 1/12 slope,and a natural sequence is formed. Then we are able to feel aesthetics of slow.

대지의 흐름을 그대로 받아들이고 건물을 땅 위에 얹힌 후 그 위에 1/12의 흐름을 입혀서 매스의 높낮이를 조절하였다. 이에 자연스럽게 형성된 지붕 천창을 통해 외부의 자연광을 내부로 끌어 들이게되며 외부와 경계지지 않은 주거 유닛이 형성된다. 각 주거공간에 형성된 발코니는 옆 벽을 막지 않아 이웃과의 소통이 가능하다.

After that, we adjusted the elevation of the mass by applying a flow of 1/12 on the hood with the building placed on the ground and try to absorb the flow of the earth as it is. Through the naturally shaped roof, natural light pulls out the exterior light to the inside, without creating a limit in the residential unit. We didn't block the walls.

TWELVE & GONGZAKSO

AIRPOCALYPSE

ABOUT AIRPOCALYPSE

airpocalypse는 공기와 대재앙의 합성어로, 공기오염으로 인한 지구 종말을 의미한다. 2013년 영국의 경제지 '파이낸셜타임스' 가 중국 베이징의 심각한 대기오염 상황을 묘사하기 위해 만들어낸 신조어로 '살인 스모그' 라고 중국에서 발생하는 스모그는 영향권이 넓고 지속 시간이 길며 오염 물질의 농도가 짙은 3대 특징을 보이고 있다며 중국 국토의 1/4에 해당하는 광범위한 지역에서 스모그가 발생해 전체 인구의 절반에 육박하는 6억명이 영향을 받았다고 말했다.

The airpocalypse, as a compound word of air and apocalypse, means the end of the earth due to air pollution. It is a term newly invented by the British financial magazine "Financial Times" in 2013 to explain the serious air pollution situation in Beijing, China. It is called "murder smog" and the Chinese smog has a wide range of influence, long duration, and highly polluted pollutants. And they reported that about 600,000 people, nearly half the population, were affected by smog in a region that is one-quarter of China's land area.

이흔희 I Lee Hunhee, 서지영 I Seo Jiyoung

KOREA SMOG FORM

서울형 스모그는 자동차 배출가스와 안개가 결합하여 나타나는 매연현상이다. 지형적으로 분지 형태여서 대기오염 물질이 확산되지 않고 자동차 배기가스가 주 오염원이라는 점에서 로스앤젤레스형 스모그의 특성을 갖는다. 주로 대도시나 공장지대의 굴뚝에서 배출되는 연기나 자동차의 배기가스 등이 바람이 불지 않는 상태가 지속되면 지표 가까이 쌓여 안개처럼 보이게 된다. 이것은 폐와 눈을 아프게 하고 호흡기질환을 악화시키며 폐가 약한 사람의 경우 사망에 이른 사례도 보고되었다.

Seoul smog is a mixture of automobile smoke and fog. It is in the basin formation and does not spread air pollutants. Thus, it is characterized by the fact that car emissions are the main cause of pollution. If the chimney or car exhaust continues in a metropolitan or factory area, it might be stacked close to the surface of the earth and looks like fog. The pollution can cause the eye and lung disease and even weaken respiratory organ. Furthermore, there are some reports that it also might cause death if one has a weak lung.

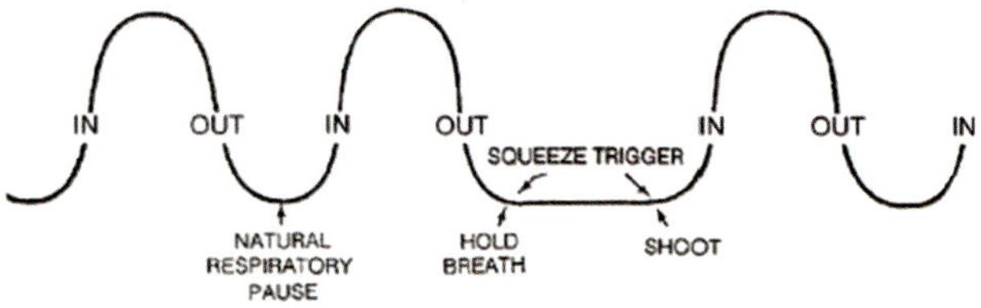

Breath control이란 호흡조절을 의미하여 숨을 들이쉬고 마시는 것을 곡선의 형태로 나타낸 것이다. 1년간 서울 주요 지역의 미세먼지 농도가 세계 보건기구의 2배로 측정되었는데, 우리는 Breath control을 이용해 서울의 대기오염을 알리는 그래프를 만든다.

Breath control is the regulation of inhaling and exhaling. We described this in a curved form. During a year, the concentration of fine dust in main Seoul areas is twice that of the World Health Organization, and we are going to make a graph using a respiratory system to report air pollution in the city.

매스 나열
Arrangement of the mass

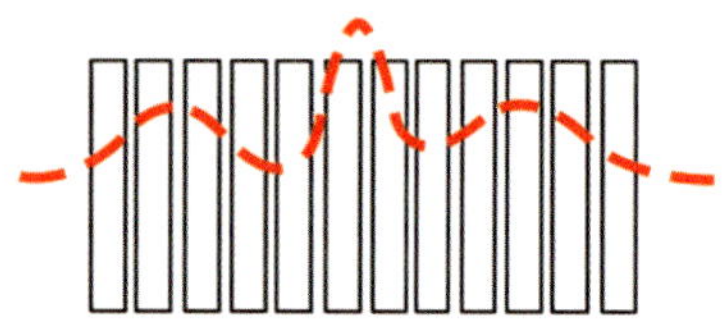
숨의 안정 – 거침 – 완화 곡선
Breath's stability- Gasp for breath- Breath's alleviation

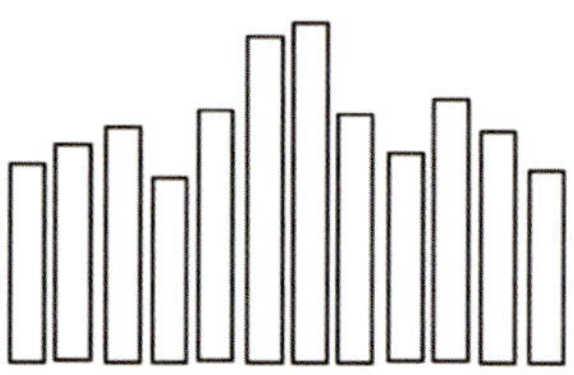
곡선에 따른 매스 변형
Transformation of the mass according to curve

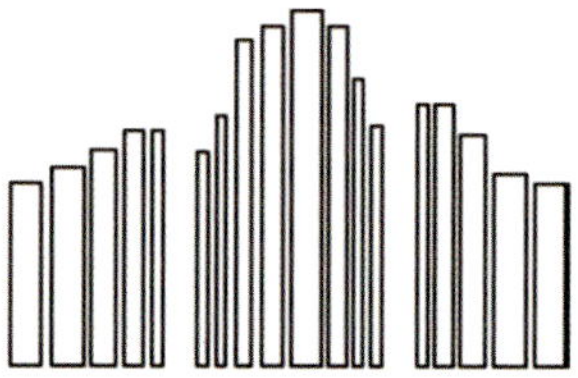
두께 조정으로 숨의 변화를 극적으로 표현
Indication of the dramatic change of the breath through the thickness control

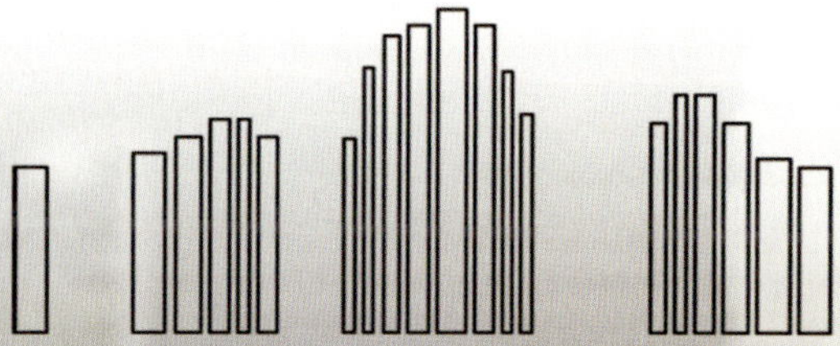
안정 – 거침– 완화 사이를 비워 공간을 구분
Separation space among the stability- roughness- alleviation

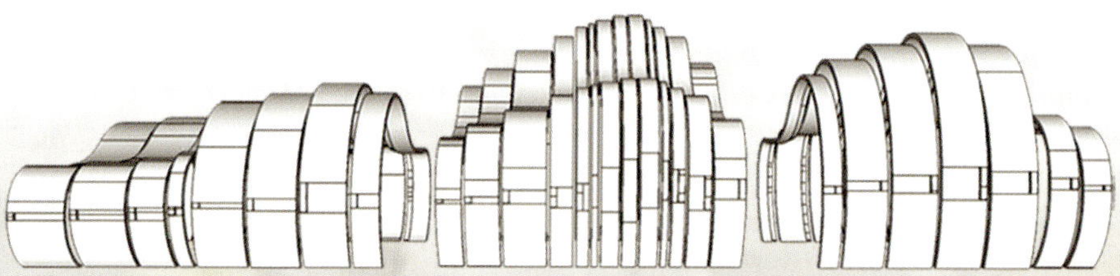
실제 모델 형태 (FACADE)
Actual shape of the model (FACADE)

BACKGROUND

우리나라의 스모그는 자동차의 배기가스, 도로 주행과정에서 발생하는 먼지들의 집합이다. 이 스모그에 노출되면 호흡기 및 심혈관계 질환의 발생 가능성이 높아지며 사망률을 증가시킨다. 이를 문제 삼아 환경보전협회는 매년 광화문광장에서 열리는 세계 환경의 날 행사에 파빌리온 '에어포칼립스' 를 세운다.

한 달 주기마다 장소를 바꿔 서울 광장, 청계광장, 광화문 광장에서 설치 가능한 포터블 파빌리온으로 여기에 엔터테인먼트적 요소를 더하여 유쾌한 체험들을 통해 스모그에 대한 위기감을 느낄 수 있도록 사람들을 유도한다.

Korea 's smog is a group of dust generated from the exhaust gas of a car and driving. If we are exposed to this type of smog, it increases the possibility of respiratory and cardiovascular diseases. Environmental Protection Association establishes an airpocalypse pavilion on the World Environment Day events that take place every year in Gwanghwamun Square.

It is a portable pavilion that can be installed at the Seoul Plaza, Cheonggye Plaza, and Gwanghwamun Square by changing the venue every month. By adding entertainment elements we will try to encourage people to feel the sense of crisis about smog through enjoyable activities.

SITE ANALYSIS

서울특별시 종로구 세종대로 172, 광화문광장 세종대왕 앞 일대 '중앙광장'

광화문광장은 서울 도심 중 미세먼지 농도 1위로 가장 공기가 안좋은 곳으로 알려져있다. 이는 서울의 중심이기 때문에 교통체증이 심하기 때문이다. 그래서 미세먼지 토론회 등이 열리기도 하였다. 광화문 광장은 바로 앞에 광화문역이 있고 사람들이 많이 몰리기 때문에 누구나 쉽게 접근 가능한 곳이기 때문에 이 곳에 파빌리온을 설치하였다.

INTERACTION DESIGN

숨을 쉬는 것에 있어 공기의 오염도는 우리 인체건강에 적지 않은 영향을 미친다. 하지만 눈에 보이지 않는 공기 오염을 인지하기는 쉽지 않다. 눈으로 볼 수 없는 공기의 흐름을 시각적 인터페이스를 통해 시각적, 촉각적으로 느끼도록 한다.

Air pollution in breathing affects our health. However, it is difficult for us to realize the air pollution that we are not able to see. Then we try to feel the flow of air in a visual and tactile way which we cannot see.

FESTIVAL INFORMATION

세계 환경의 날 (World Environmental Day)

세계 환경의 날 행사는 매년 6월 5일에 열린다. 1972년 6월 스웨덴 스톡홀름에서 열린 '유엔인간환경회의' 개최 시 제정, 그 해 UN총회에서 채택되었다. 우리나라는 1996년부터 6월 5일을 법정기념일인 '환경의 날' 로 제정했으며, 1997년엔 서울에서 UNEP주최의 '세계 환경의 날' 행사를 개최하기도 하였다.

World Environmental Day

The World Environment Day event will be held on June 5th every year. It was enacted at the United Nations Conference on the Human Environment, which was held in Stockholm, Sweden in June 1972 and adopted at the UN General Assembly that year. In Korea, we designated the legal anniversary ' Environment Day' on June 6th,1996 and in 1997, we also held UNEP's World Environment Day in Seoul.

Gwanghwamun Plaza, Sejong-daero 172, Jongno-gu, Seoul, Korea

Gwanghwamun Plaza is the best dust concentration area in Seoul and is also known for its high air pollution. Because it is the center the of Seoul and has the heavy traffic jam. In Gwanghwamun Plaza, there is Gwanghwamun station. Hence, it is easy for everyone to approach.

INTERACTION DESIGN

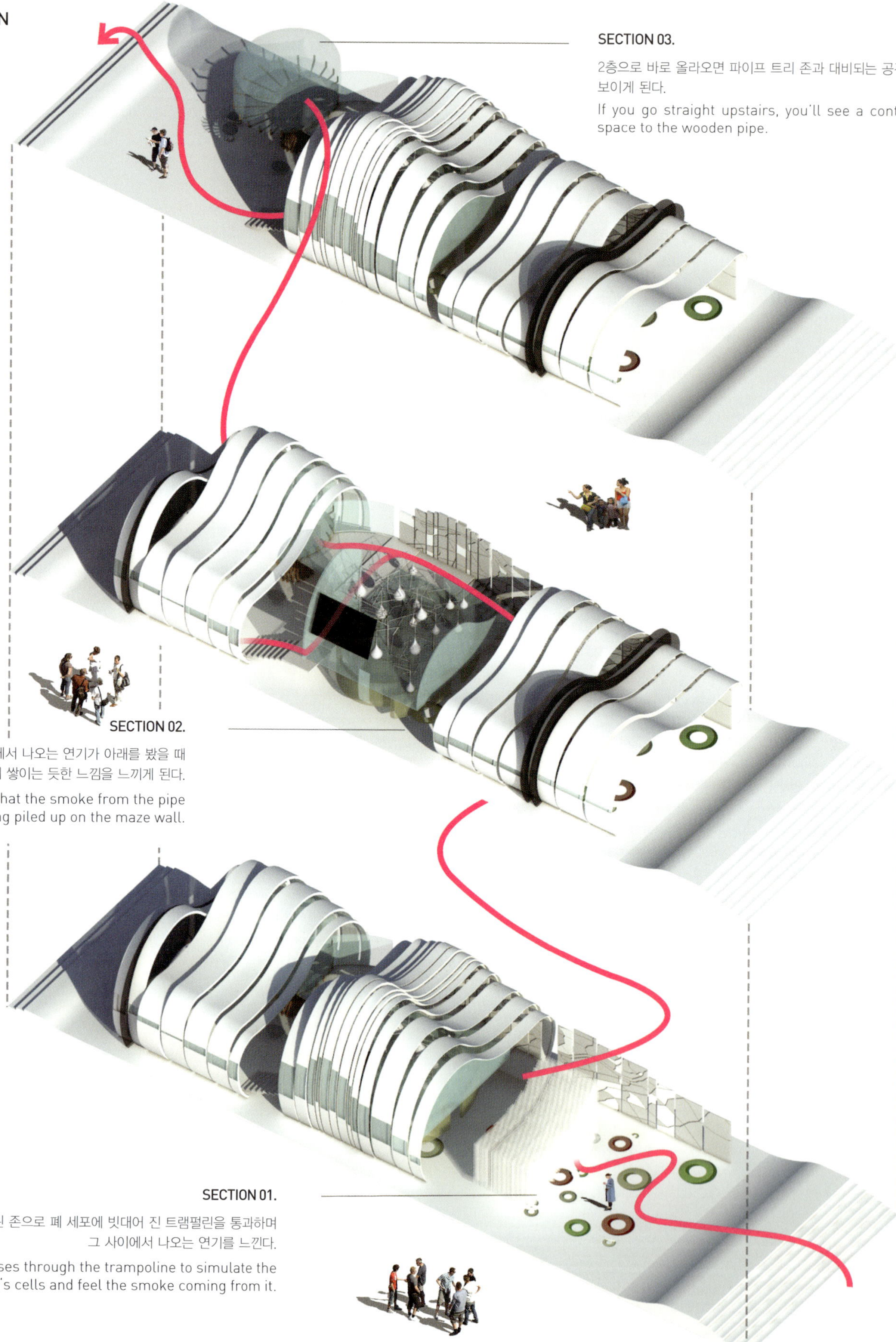

SECTION 03.

2층으로 바로 올라오면 파이프 트리 존과 대비되는 공간이 보이게 된다.

If you go straight upstairs, you'll see a contrasting space to the wooden pipe.

SECTION 02.

2층도 파이프 트리 사이에서 나오는 연기가 아래를 봤을 때 보이는 미로벽에 쌓이는 듯한 느낌을 느끼게 된다.

And you would feel that the smoke from the pipe tree is being piled up on the maze wall.

SECTION 01.

트램펄린 존으로 폐 세포에 빗대어 진 트램펄린을 통과하며 그 사이에서 나오는 연기를 느낀다.

It passes through the trampoline to simulate the lung's cells and feel the smoke coming from it.

PLAN

2F

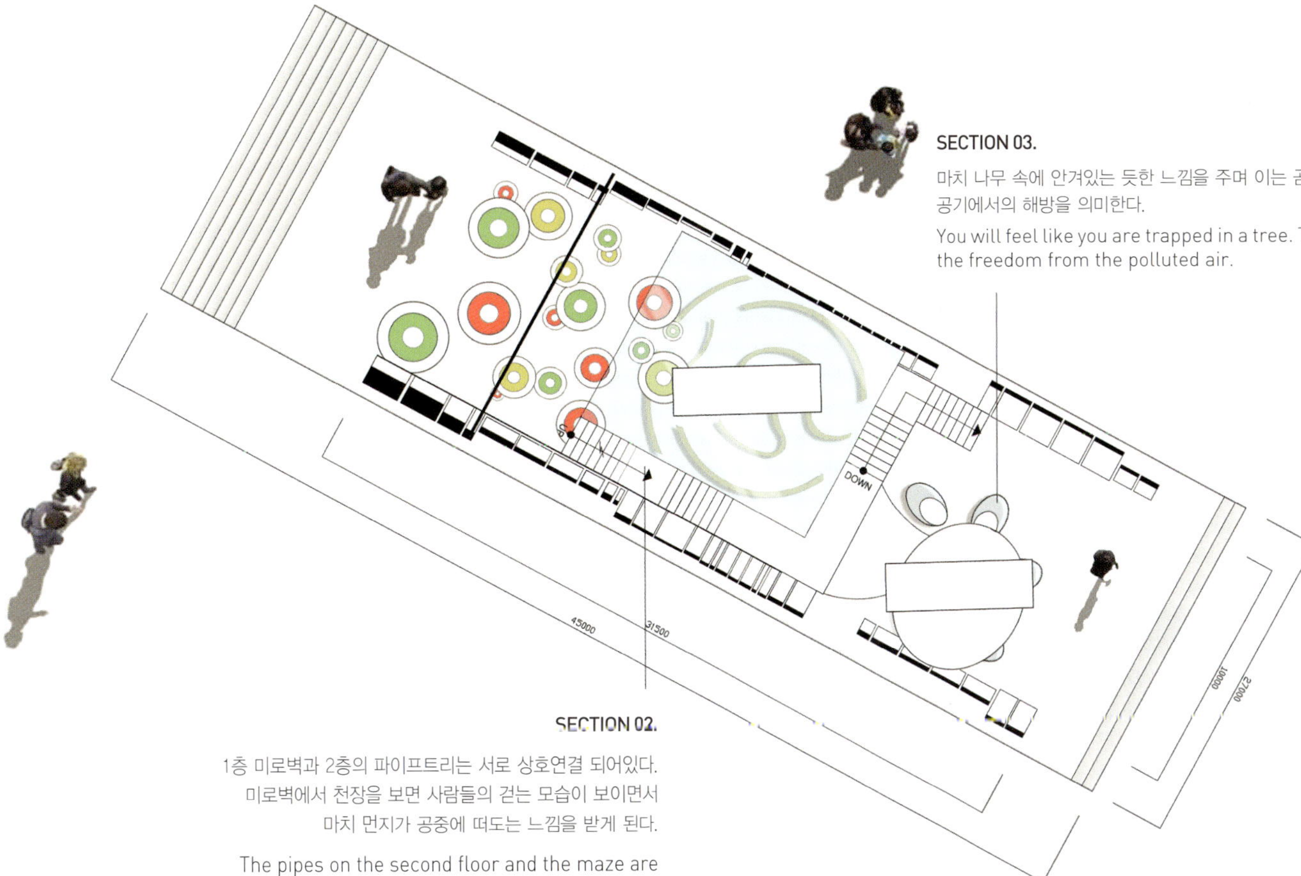

SECTION 03.

마치 나무 속에 안겨있는 듯한 느낌을 주며 이는 곧 오염된 공기에서의 해방을 의미한다.

You will feel like you are trapped in a tree. This means the freedom from the polluted air.

SECTION 02.

1층 미로벽과 2층의 파이프트리는 서로 상호연결 되어있다. 미로벽에서 천장을 보면 사람들의 걷는 모습이 보이면서 마치 먼지가 공중에 떠도는 느낌을 받게 된다.

The pipes on the second floor and the maze are connected. When you look at the ceiling from a labyrinthine wall, you can see people's footsteps and feel the dust in the air.

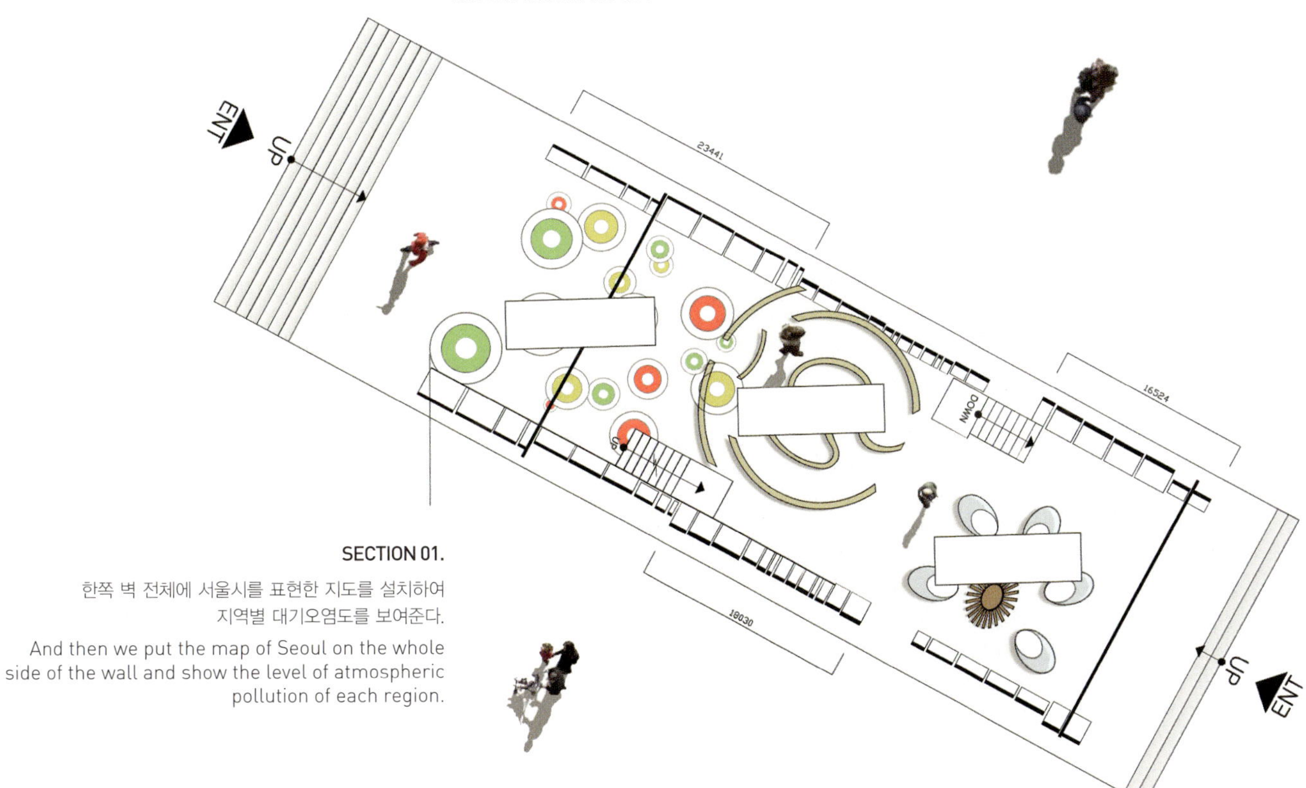

SECTION 01.

한쪽 벽 전체에 서울시를 표현한 지도를 설치하여 지역별 대기오염도를 보여준다.

And then we put the map of Seoul on the whole side of the wall and show the level of atmospheric pollution of each region.

SYMBOL

폐의 세포를 트램펄린으로 표현해 사람들이 활동하면서 발생되는 이산화탄소를 자동차 배기가스에 빗대어 사람들의 활동량이 높아질수록 트램펄린에서 나오는 빛이 선명해지며 연기가 나오는 것으로 배기가스로 인해 오염된 현재를 표현한다.

By expressing lung cells as trampolines, carbon dioxide produced by people's activities is added to automobile exhaust. As the activity of a person increases, the light of the trampoline becomes clearer and the smoke comes out to express the polluted situation due to the exhaust gas.

트램펄린에서 다양한 빛을 나오게 함으로써 혼란스런 느낌을 더한다. 빛의 선명도와 주변에 반사되는 빛들로 인해 분위기가 달라지며 사람들의 이목을 집중시킨다.

By letting various light from the trampoline, we add the feeling of confusion. Due to the reflected light, the atmosphere is being changed and catch peoples attention.

PERFORMANCE

낮에는 사람의 온도에 의해 반응하며 밤에는 대기의 오염도에 따라 빛이 달리 나타난다. 위험도가 높으면 적색등이 켜지며 좋은 날에는 다양한 빛이 연출되며 마치 생명체가 살아 움직이는 듯한 느낌을 준다. 파빌리온 곳곳에서 연기가 나오면서 마치 생명체가 숨을 내뿜는 듯한 느낌을 준다.

Section 1 에서 만들어진 연기로 인해 스모그가 점점 쌓이는 것을 미로처럼 벽을 만들어 혼란을 극대화 한다. 구불거리는 미로 벽은 유리로 이루어져 안에 모래가 가득 쌓여있고 이동하는 사람들은 그것을 보며 스모그로 인해 앞이 가로막힌 느낌을 받는다.

Smoke in Section one creates a maze wall for maximum confusion. The rotating maze walls were made of glass and filled with sand, and people who were moving will feel like they are blocked by smog.

폭이 좁고 구불구불한 미로 벽을 지나면서 답답함을 느끼게 한다. 트램펄린 위에서 활동한 뒤 바로 펼쳐지는 좁고 답답한 시퀀스에 사람들은 불편함을 느끼게 된다. 마치 스모그로 인한 답답함을 느끼는 것과 같다.

It is stuffy when people pass through the walls of narrow, crooked maze. People feel uncomfortable with the narrow and stuffy space after the experience on the trampoline. It is like smog makes people feel the space more stuffy.

It reacts by temperatures of the people on the daytime and by the atmospheric pollution level at night. Smoke is coming out from the pavilion makes people feel like living things are exhaling.

SYMBOL

대기오염이 과부화되어 서울시의 폐기능을 하는 나무가 정화시켜주는 기능을 잃어가는 모습을 녹슨 파이프에서 연기가 비정상적으로 나오는 것으로 연출한다. 그리고 tv를 설치해 스모그에 갇힌 광화문을 연속적으로 보여주며 현재 상황을 인식시킨다.

We then described air pollution, which is over-supplied as a lost function of the lungs of the city, making it seem like dirty smoke coming out of a rusty pipe. We would also continuously shows the smog in Gwanghwamun by showing them through a TV. We will install a TV for them to recognize the current situation.

2층으로 올라와 나무 안으로 진입할 수 있는 시퀀스가 전개된다. 마치 빨려들어 가는 듯한 형태의 다리를 지나 유리막 안으로 들어가는 것은 마치 나무 속으로 들어가는 듯한 느낌을 받게 된다.

There is another progression, where we can enter into trees on the second floor. Passing through bridges that look like people are being sucked in, inside the glass membrane feels like they're going into a tree.

사람들이 앞의 전시들로 스모그를 인지하고 관심을 갖게 되면서 대기오염의 개선을 인식해 나무를 심는 행위를 나무에 피부를 접촉하는 행위로 비유한다. 그로인해 나무는 성장하고 스모그에 갇혀있던 서울시에 신선한 공기가 증가하게 된다.

As people become aware of smog and get interested in the exhibition, they compare tree planting to trees to improve air pollution. As a result, trees will be growing and there will be fresh clean air in Seoul, which is stuck in smog.

나무 형상을 한 매스 주변으로 같은 형태의 작은 나무들을 설치한다. 나무 위에 얹어져 있는 판을 누르게 되면 연기가 나온다. 사람들이 대기오염에 대해 관심을 갖는 행위를 작은 나무에 접촉하는 것으로 비유했다.

Install small trees of the same shape around a mass of wood. Smog will come out it people push a board on the tree. People's interest in air pollution will be compared by touching those small trees.

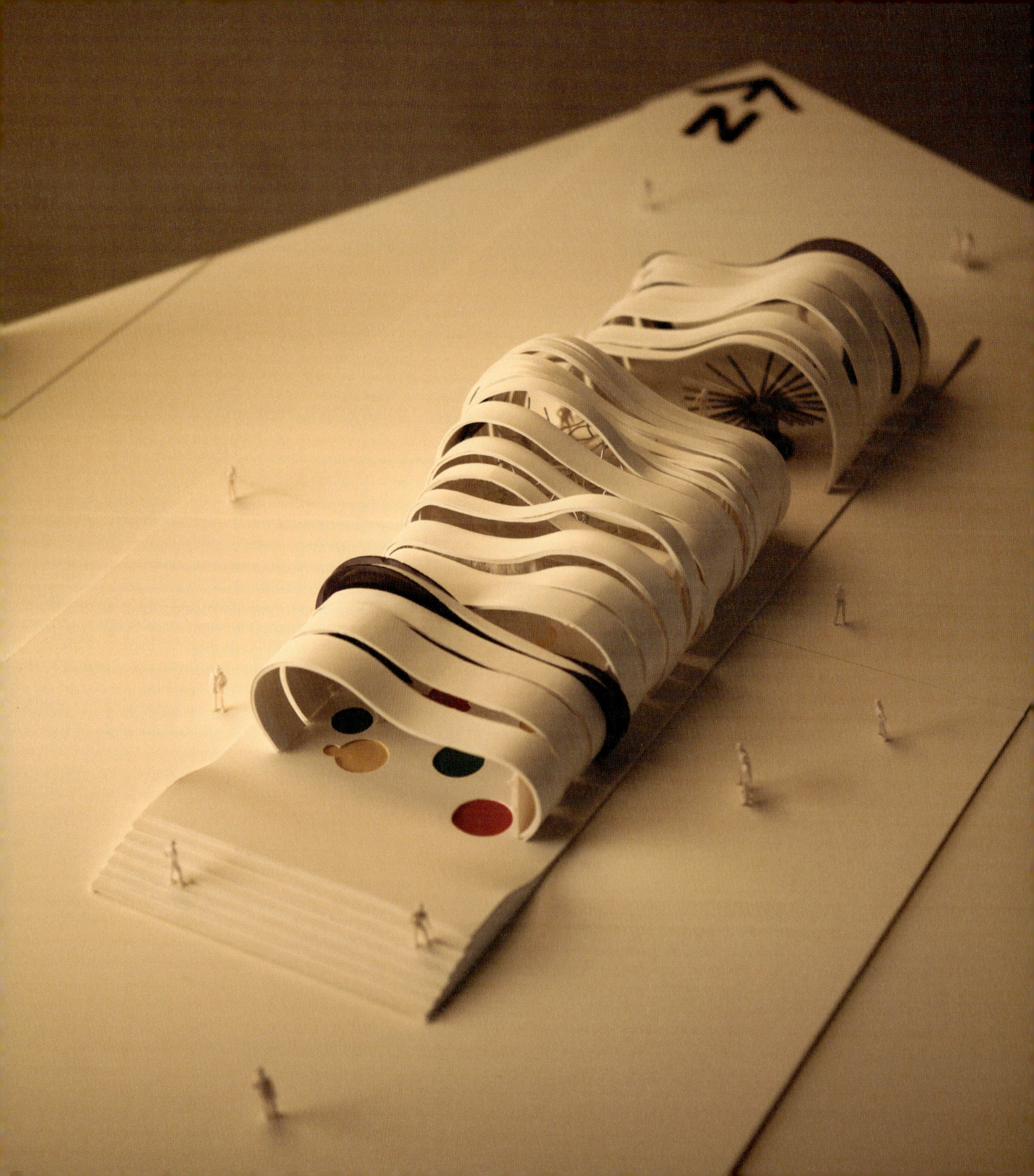

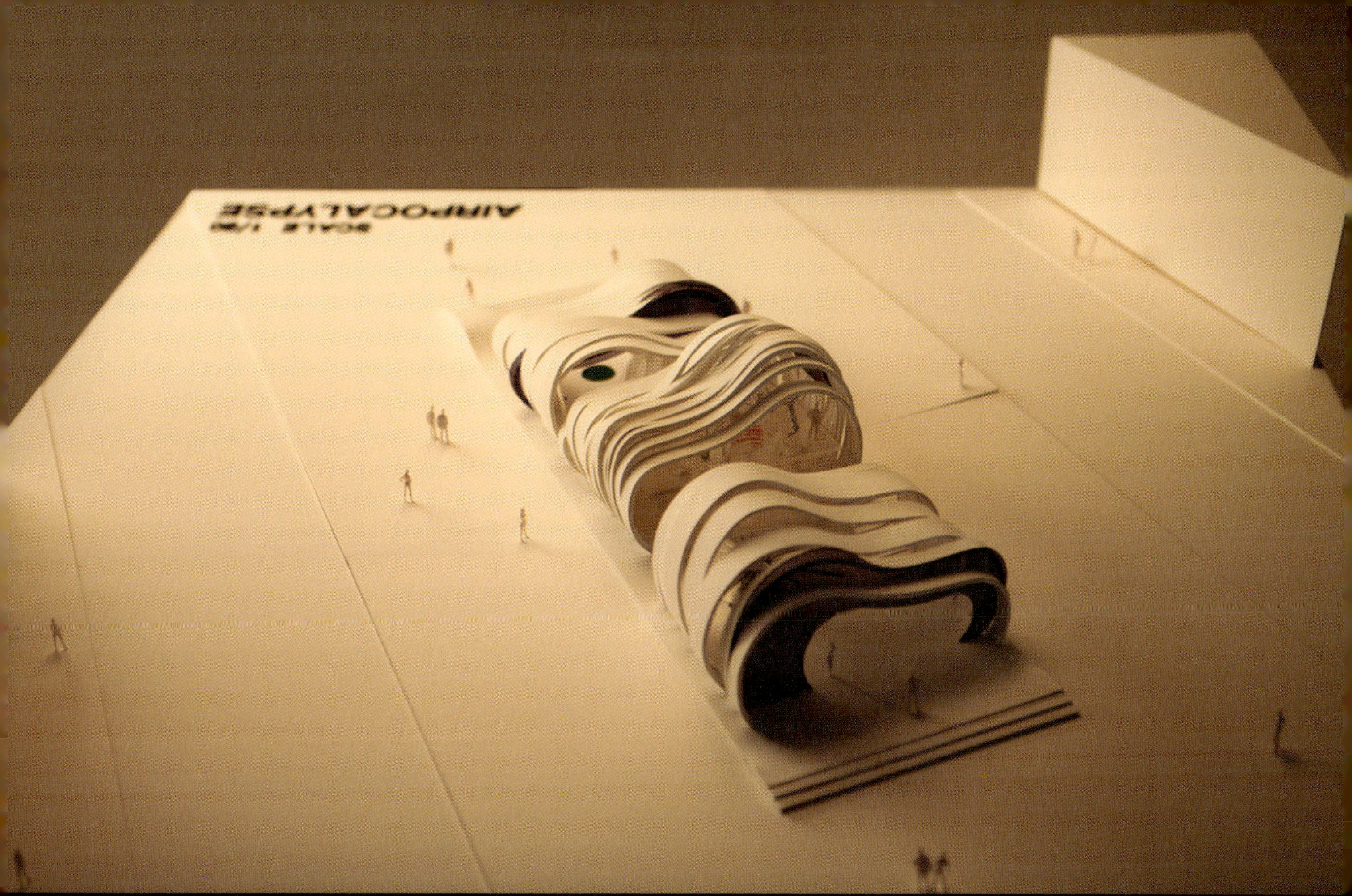
AIRPOCALYPSE

정진아 | Jung Jina, 홍세미 | Hong Semi

Star of the Sea

바다의 별이 되어 조난을 막다
Becoming a star of the sea to prevent disaster

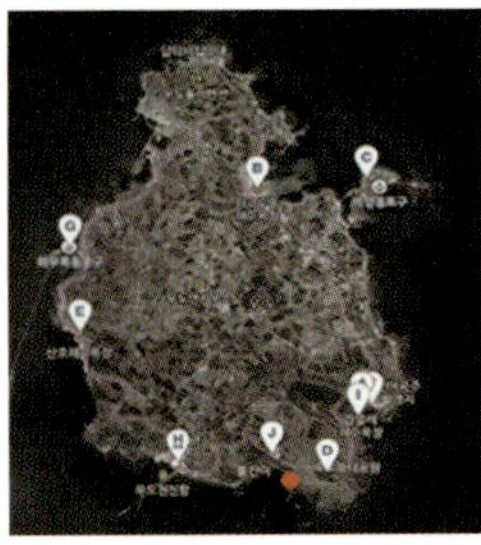
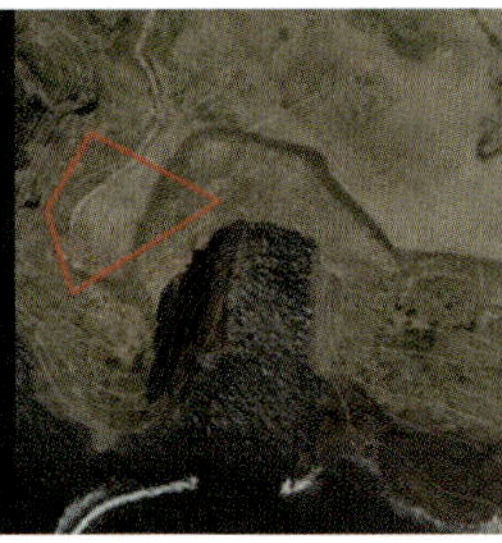

SITE

제주시 우도면 연평리 산 26

Jeollanam-do Yeonpyeongri Mountain 26

SITE ANALYSIS

지면으로부터 튀어나온 곶이 형성된 지형으로 지대가 높기 때문에 등대의 광원을 쏠 때 멀리까지 닿아 선박에 위험 경고 표시 및 항로 결정에 용이하다. 우도 천진항과 가까운 거리에 위치하여 선박의 유입이 매우 잦은 곳으로 재난에 대비한 등대가 필요한 곳이다. 우도 파출소와도 가까워 조난 시 상황 전달이 용이하다.

Since it is a land where the cape protrudes from the ground is high enough, it is easy to indicate the warning sign and determinate route on the ship when the light of the lighthouse is reached far away when it shoots the light.

Udo - It is located near Udotjin Port, which is very famous for carrying ships to the lighthouse in case of a disaster. Moreover, It is close to the Udo police office, so it is more convenient to report the situation in case of emergency disaster.

BACKGROUND

요즘 미디어 매체에서 바다에서의 안타까운 조난 사고를 어렵지 않게 들을 수 있다. 무역 선박, 군 선박뿐만 아니라 바다 낚시 인구가 매해 증가함에 따라 낚시 어선의 사고도 2017년을 기준으로 3년새 170% 증가하였다.

조난에 대한 안전 불감증이 커진 현재, 강렬한 빛으로 선박에게 육지의 소재, 원근, 위험한 곳 등을 명시하여 조난을 방지하는 등대를 만들어 우도 주민과 선박의 안전을 지키는 공간을 제안하고자 한다.

These days, many sad accidents that have happened in the sea are being heard in media. As of 2017, even fishing boat accidents have increased up to 170% in three years, at the same time the numbers of fishermen as well as trade ships and military ships have been extremely increased.

We are planning to set up a lighthouse to protect residents and ships from disaster by indicating every exact location, view, and danger areas with stronger lights.

PURPOSE

섬 속의 섬이라 불리는 우도는 제주도를 찾는 사람이라면 반드시 들린다고 할 정도이며, 지난해 관광객은 무려 300만 명에 육박했다. 그러나 우도의 실제로 거주하는 우도 주민은 1800명이 채 되지 않는다. 이러한 우도 주민들을 위해 우도 바다에서의 재난을 막아 안전을 지키는 등대의 역할을 하면서도 우도 주민이 자유롭게 와서 휴식을 취하고, 책도 읽으며 사회 공동체 망을 형성하도록 하는 커뮤니티 문화 공간을 제안한다.

Udo so-called the island in the island, reached 3 million tourists last year. Almost every tourist who visited the Jeju island must be absolutely visited Udo as well. Nonetheless, there are less than 1800 residents in Udo.

For these residents, the lighthouse is not only for protection from disaster but also plays a role as a community space to build up a social network such as a rest area and as a library as well.

PURPOSE

"Constellation : 별자리"

예로부터 사람들은 길을 잃었을 때 별자리를 보고 방향을 잡아 길을 찾곤 했다. 등대 역시 바다에서의 길라잡이라는 점에서 별자리와 동일한 역할을 한다. 이러한 동일한 나침반과도 같은 역할을 하는 별자리를 등대에 접목해 공간에 표현한다. 별이 빛나는 부분을 공간 곳곳에 대입하여 빛의 확산과 흡수로 밤낮에 따라 공간의 반전을 느끼게 하는 등대 공간을 만들고자 한다.

From time immemorial, people used to look at constellations and find directions when they were lost. The lighthouse also plays the same role as a constellation in that it is a guide in the sea. These constellations, which play the same role as the same compass, are incorporated into the lighthouse and expressed in space. We want to create a lighthouse space where stars are shining in space to spread the light and absorb light day and night.

DESIGN PROCESS

별자리 중 한 점에서 시작했다가 두 갈래로 갈리고 다시 한 점으로 만나 쭉 뻗는 작은 곰 자리의 형태를 동선에 적용

Applying a shape of a little bear to the flow starting with one of the constellations. After that split it into two ways and then let them be combined again.

입구에서 두 슬라브가 갈라져 한 슬라브는 오직 내부로만 갈 수 있으며 또 다른 슬라브는 외부로만 갈 수 있어서 3층까지는 두 슬라브가 만날 수 없는 동선

The moving line with two slabs at the entrance is unable to meet each other till they reach to the third floor. One slab can only go internally, the other can only go externally.

만난 두 슬라브가 한 개의 슬라브로 합쳐져 다시 위층으로 뻗어가는 동선

A line, which combines into one slab again and moves upstairs.

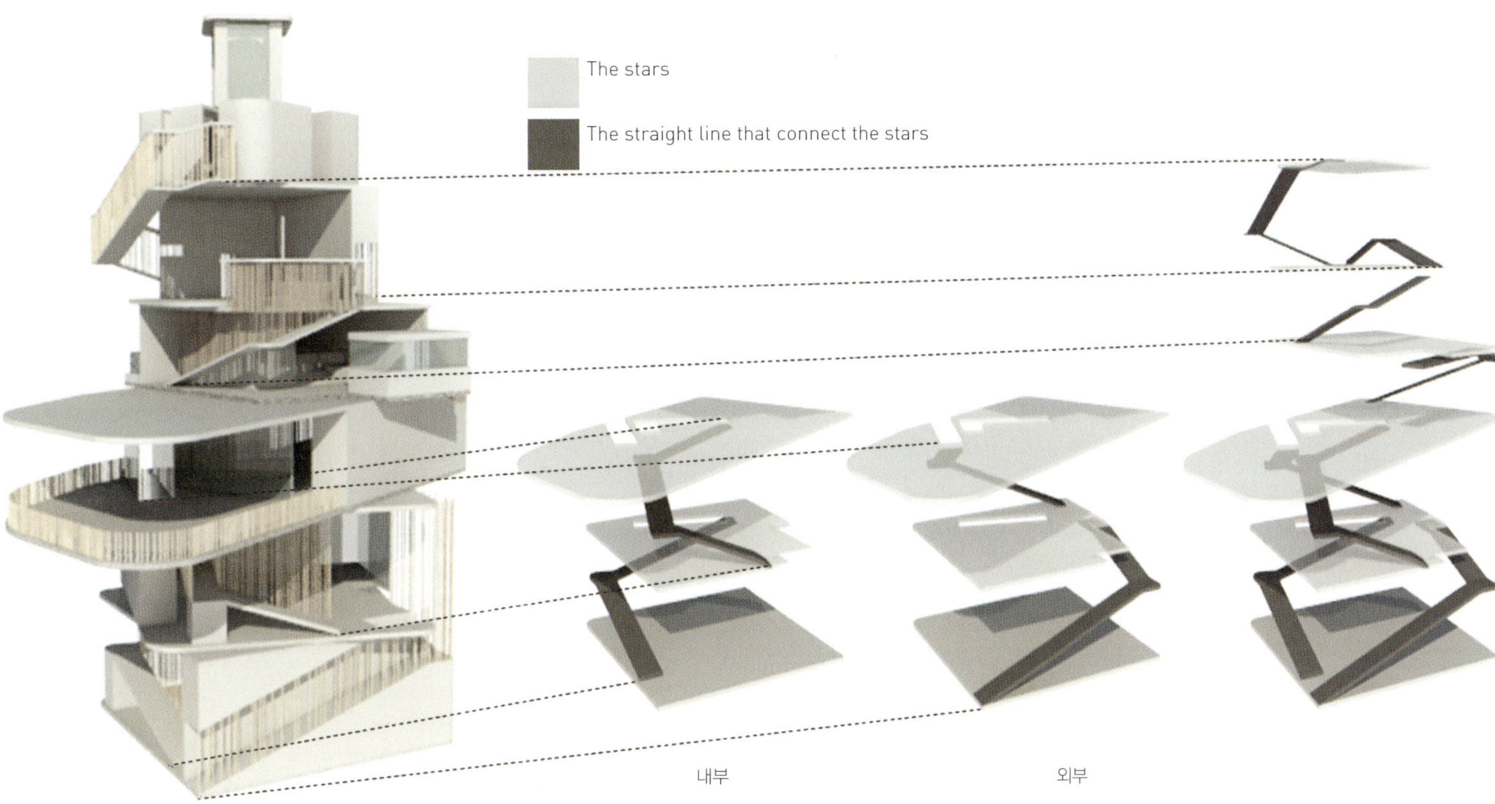

PERSPECTIVE

5th Floor

5층은 등대지기들이 등대의 본 역할인 바다에서의 재난을 방지하게 하는 오피스 공간이다. 오피스 테이블은 겹쳐진 우드판 사선으로 휘어지고 삽입되고 튀어나오게 하여 다소 지루해질 수 있는 오피스 공간에 재미를 불어넣고자 하였다.

The fifth floor is an office. The lighthouse keepers are responsible for preventing the main role of the lighthouse disaster in the sea. We bent the wood plate on top of the office table diagonally, and then made it protruded to make the office lively that might be rather boring.

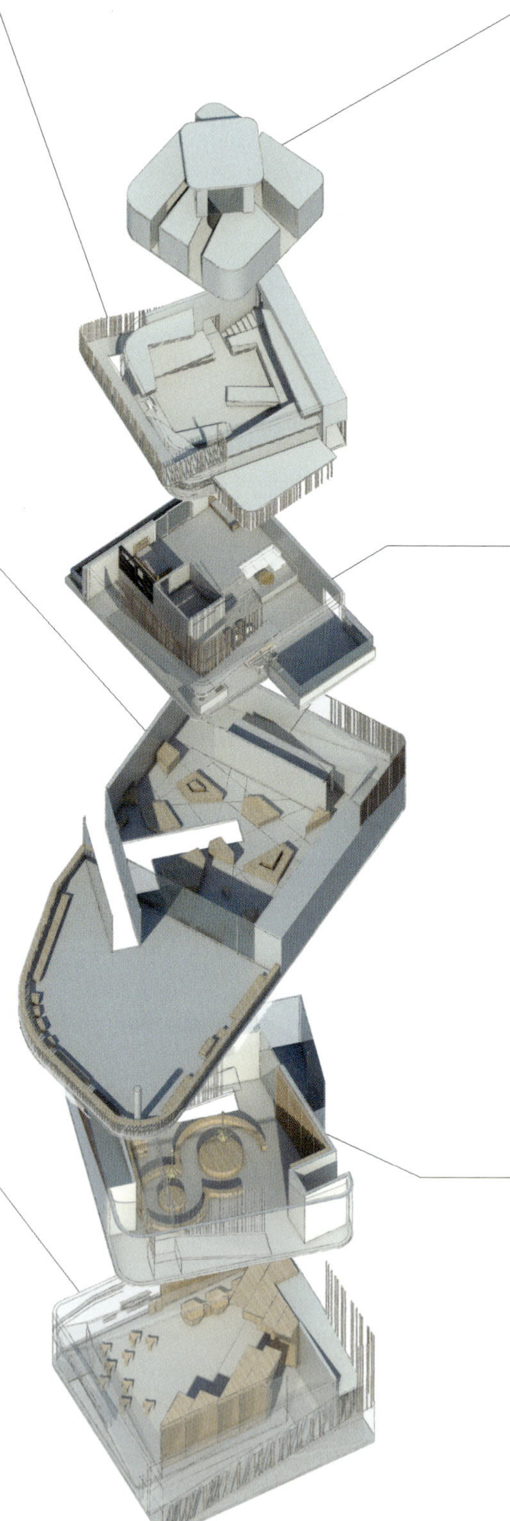

3rd Floor

3층은 외부 슬로프와 내부 슬로프가 만나는 곳으로, 만남의 광장과도 같은 커뮤니티 공간이며 전망대이다.
슬라브를 길게 빼고 필로티를 두어 공간의 반은 외부, 반은 외부로 설정하였다.
천장에 기하학적으로 불규칙하게 튀어나온 도형들 사이에 조명을 넣어 별자리를 형상화 하였다.

The third floor is the area where the external slope meets the internal slope, which is the same as the conference area and observatory. We set up the slab long and place a column out in half space. Then we embodied constellation by putting lights between the ceiling that were geometrically and irregularly stuck out.

1st Floor

1층은 안내데스크와 로비가 있다. 눈높이에 맞춰 창을 뚫어 창 밖으로 보이는 우도 전경이 마치 사진을 보는 것 같은 픽처프레임 효과를 주었다. 로비로 들어서는 곳과 안내데스크는 동굴을 연상케 하는 벽을 두었고 그 벽 사이로 뿜어져 나오는 빛이 공간 안에 긴장감을 준다.

The lobby and the information desk in on the first floor. We gave av picture frame effect depending on the eye level as if it were looking at a photo of the Udo's view out of the window. Lobby and information area have walls that look like caves, and the light between the walls keeps the space edgy.

The Tower

5층에서 등탑으로 가는 슬라브는 벽에서 툭 튀어나와 형성되었으며, 등탑 공간은 보이드와 솔리드한 공간으로 두어 틈새 사이의 빛의 방출과 흡수로 인한 반전있는 매스감을 느낄 수 있다.

Slabs from the fifth floor to the top are formed by protruded walls and loop spaces and the lantern space is made of voids and solid space, so visitors can feel the inverted mass feeling due to the emission and absorption of light between the gaps.

4th Floor

4층부터는 슬로프가 하나로 만나 올라가는 곳으로, 일반인들은 갈 수 없고 등대에서 일하는 등대지기만을 위한 곳이다.
4층은 등대지기의 숙소로 주거 공간이다. 3200mm의 높은 천장 공간을 바닥에 층고가 다르게 레벨 변화를 주어 활용하였다.

From the fourth floor, the slope meets and going up as one, and is managed only by the lighthouse staffs and is not open to the public. The fourth floor is a residential area of lighthouse staffs. We then utilized and gave it some change to the ground of a high ceiling of 3200mm through the different height of the floor.

2nd Floor

2층은 우도 주민들이 책을 볼 수 있는 문화공간으로 설정하였으며, 크게 뚫린 창을 통해 내부의 슬로프와 밖의 전경을 볼 수 있어 공간의 개방감을 주었다. 슬로프의 난간 대신 얇고 긴 루버를 주어 별들이 떨어지는 듯한 몽환적인 느낌을 주고자 했다.

We set up the second building as a cultural space where residents can reading books. Through the big window, they can even see the outside view and slope. Also, we focused on the openness. Instead of a balustrade, we attached thin and long louvers to give an effect as the stars are dropping.

PLAN VIEW

1st Floor

2nd Floor

3rd Floor

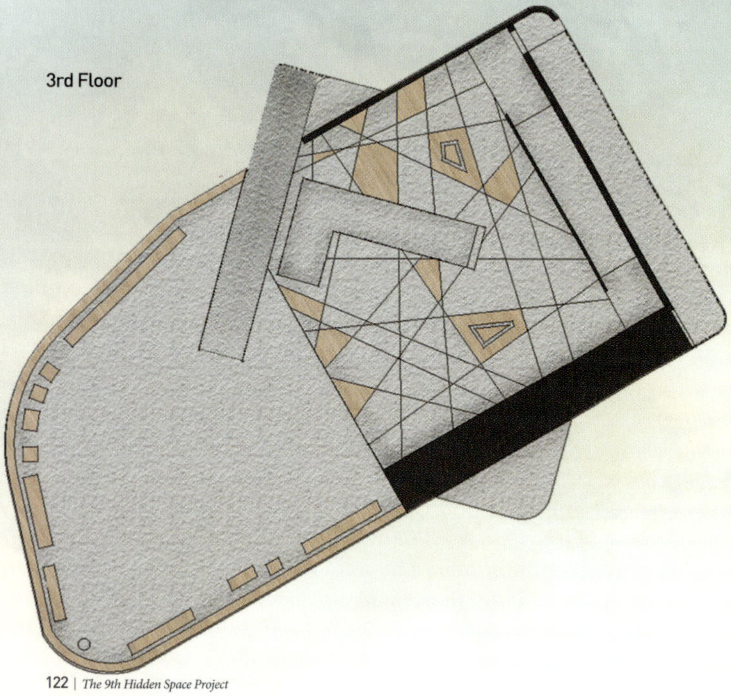

4th Floor

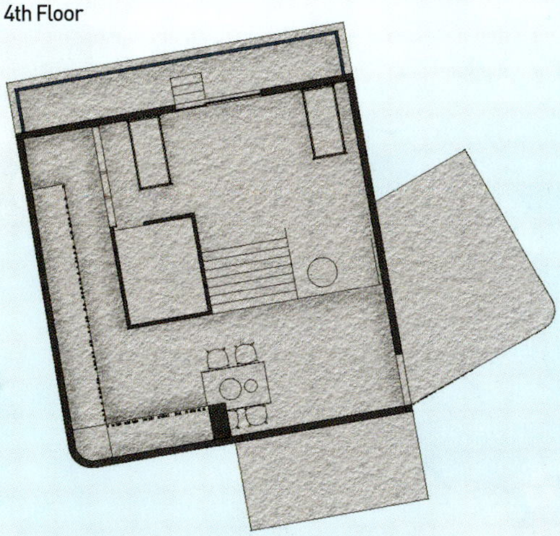

5th Floor

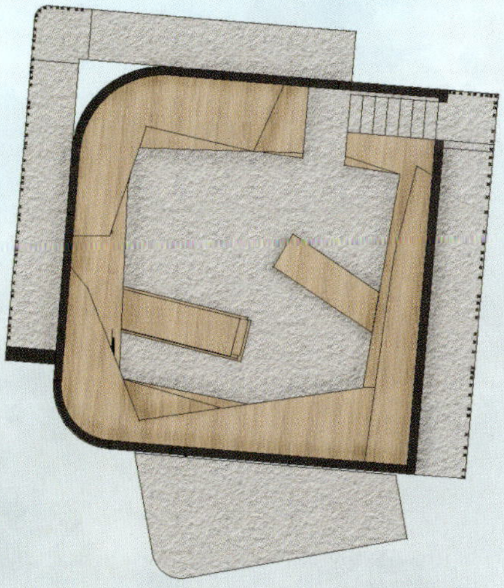

The Tower

기존의 등대하면 떠오르는 원형의 기다란 형태의 틀을 깨고자 1층의 중심을 축으로 각 층별로 15, 20, 30도씩 틀어 배치하고, 층별로 공간의 크기와 층고의 높이를 다르게 하여 매스의 높고 낮은 불규칙한 리듬감을 주었다.

We arranged from the base of the first floor at 15, 20, and 30 degrees to break the traditional, circular paterns of the light-house, and set the level of space and height, which were different by layer, giving the mass a high and low irregular rhythm.

1st Floor_Information

2nd Floor_Library

3rd Floor_Observatory & Community Space

오세정 I Oh sejeong, 이수민 I Lee su meen

Stairs with space

계단을 이용한 소통공간의 지하벙커를 만들다.

Create an underground bunker of communication space using stairs.

BACKGROUND

대한민국은 전쟁으로부터 안전한 국가가 아니다. 실질적으로는 종전이라고 하지만 아직 6.25전쟁이 휴전한 상태이다. 또한 북한과 미국의 정치적 갈등이 나날이 심화 되고 있는 현 상황에서 더 이상 안심 할 수 없다는 생각이 들었다. 앞으로 전쟁이 일어난다면 핵전쟁일 가능성이 높다. 하지만 사람들은 아직 일어나지 않은 핵전쟁에 대한 경각심이 부족하고 대피공간에 대한 관심 또한 부족하다. 그러나 현시점에 서울에 핵이 폭발한다면 적절히 대피할 수 있는 공간이 마련되어있지 않다. 또한 서울과 수도권 주변의 지역들은 방사능으로부터 자유로울 수 없다. 이러한 문제를 해결하기 위해 대피가 가능하고 수년간의 생활을 유지 할 수 있는 지하벙커를 설계하고자 한다.

South Korea is not a safe country from war. Some say the war has ended already, but in fact, the 6.25 war is still in a state of truce. I also felt that I could no longer be relieved in the current situation of conflicts between North Korea and the United States. If there is going to be a war, it is highly likely that it will be a nuclear war. However, most of the people have lack of awareness about nuclear war that has not happened yet, and there is also a lack of concern about evacuation space. However, if nuclear weapons
explode in Seoul at this point, there is no room to evacuate. Also, areas around Seoul and its metropolitan areas are not free from radiation. To solve these problems, we came up with the idea to install an underground bunker. It is to provide shelter and support and maintain for many years of living.

SITE ANALYSIS

설치구역은 초, 중, 고 등의 운동장으로 설정했다. 학교는 상징성을 띄고 있어서 대피 상황에찾아 오기 쉽고 넓은 운동장은 커다란 지하벙커를 수용하기에 매우 적절하다고 생각했다. 또한 학교의 특성상 주변의 건물이 크게높지 않기 때문에 건물이 무너졌을 경우 매몰 될 가능성이 적기 때문에 대피 이후 빠져 나올 때의 상황까지 고려하였다. 대표지역을 인천광역시 강화군 길상면에 딸린 작은 섬 동검도로 설정했다. 동검도는 면적1.61㎢에 인구수가 약130여명의 작은 섬이다. 이곳을 위치로 설정한 이유는 섬이라는 특성상 대피에 더 어려움이 있을 것이라는 생각이 들었다. 또한 수도권 지역의 섬이기 때문에 핵폭발과 방사능으로부터 자유롭지 않을 것이라는 생각이들어 이곳을 위치로 선정했다.

I thought playgrounds such as an elementary, middle, and high school as out installation area. This is because schools are symbolic and easy to get to, and
I thought the large playgrounds are appropriate to accommodate the large underground bunkers. Moreover, due to the feature of the school, the surrounding buildings are not very high, so when the building collapses, it is less likely to be buried.I even considered the situation when I got out of the evacuation. I chose the main area as a small island called Donggumdo in the Gangsang-myeon, Ganghwa-gun, Incheon. This Island is a small island with an area of 1.61㎢ and a population of 130 only. The reason I set this island as the main area was that I felt it would be more difficult to evacuate due to the feature of the island. Besides, it is an island in the metropolitan area, I thought that it would not be safe from nuclear explosion and radiation.

TARGET

대상은 동검도의 거주자 전체로 잡았다.
섬이기 때문에 대피하기 난감한 위치에 있는 대상일 뿐만 아니라 서울과 가까이 위치해 있어 그 피해가 예상 가능하므로 동검도 지역 주민들을 대상으로 잡게 되었다.

I chose all the citizens as the main target. As it is an island and located close to Seoul,the damage is foreseeable, and Donggum Island has also been targeted on local residents.

STORY

지하벙커를 주제로 선정한 후 어떤 단점이 있는가에 대해 생각해보았다. 여러 문제들이 있겠지만 그 중 가장 큰 문제는 공동체 생활에서 오는 문제들이었다. 각자의 삶을 살던 사람들이 어느 날 재난으로 인해 한 곳에 모여 살게 된 것이다. 갈등이 일어날 수밖에 없다는 생각이 들었다. 이러한 문제를 극복하기위해 생각하던 중 판옵티콘과 계단을 주제로 설정했다.

I have thought about the weaknesses after choosing an underground bunker as my topic. There must be many problems, among them, I thought the biggest problem will be community life. When people lived their lives were gathered together in one place due to a disaster, I thought that there must be a conflict of course.
To overcome this kind of problem, I set up panopticon and staircase as my main point.

CONCEPT

:판옵티콘 PANOPTICON

공동체 생활에서는 규칙과 질서가 필수적이라고 할 수 있다. 하지만 보이지 않는 곳에서는 질서와 규칙이 무너지기도 한다. 그렇다면 '보이지 않는 곳이 없다면 어떨까' 라는 생각이 들었다. 그로 인해 떠올린 것이 바로 판옵티콘이다.판옵티콘은 원래 죄수를 효과적으로 감시하기 위한 방법의 일종이다.이 방법을 지하벙커에 변형시켜 적용하고자 했다.감시자와 피감시자를 따로 두지 않고 대피자들이 서로서로를감시하고 견제 하는 체계를 만들었다. 이 행위를 통해 스스로의 안전함을 찾을 수 있다. 또한 감시가 아닌 관찰의 개념으로볼 수도 있으며 관찰은 갑작스러운 공동체 생활 속의 낯선 사람들을 더 잘 파악 할 수 있도록 해 수년간 함께 지내야 할지도 모를 사람들과의 관계에 있어 적응 할 수 있도록 돕는 역할을 한다.

Rules and regulations and order are the essential points in community life. But rules are easy to be ignored and disregarded in invisible. Then I thought, 'What if I could find nowhere to hide? The reason I came up with the panopticon is that panopticon was originally a method of monitoring prisoners effectively. I tried to apply this method to an underground bunker. To make it clear, without separating the monitors from the observers, the evacuees formed a system to monitor and check each other. With this method, people can find themselves safe. It can also be seen as a concept of observation rather than surveillance. Furthermore, observation helps people to understand better strangers in a sudden community life and adapts to their relationships with those who have been together for years.

:계단 STAIR

판옵티콘의 개념을 가장 잘 나타낼 수 있는 구조가 계단이라고 생각했다. 계단은 보통 층과 층의 사이를 이어주는역할의 구조물이지만 이 지하벙커의 계단은 피라네지를 연상시킨다. 피라네지의 계단은 복잡함과 반복, 연속됨을 특징으로갖고 있는데 이 지하벙커의 계단에 적합하다. 단순히 층과 층을 이어주는 개념이 아니라 모든 공간과 공간을 이어주는 역할을 한다. 또한 계단의 폭을 넓혀 계단 자체가 공간이 되기도 하고 층과 층의 연결이 거대한 계단처럼 보이기도한다. 이렇게오픈된 공간인 계단 자체를 공간으로 활용함으로써 자연스럽게 사람들과의 만남이 가능하고 관찰 또한 쉽게이루어질 수 있다.

I thought the staircase was the best thing to describe the structure of panopticon. Stairs are usually a structure that connects floors to floors. However, the staircase in this underground bunker reminds me of Piranesi. The characteristic of Piranesi is complexity and repetition, and continuation and I believe that these traits are suitable to the underground bunker. It is not just a concept of connecting floor to the floor, but it connects all spaces and spaces. Further, the staircase becomes a space itself, and the connection between the floor and floor may seem like a huge staircase. By using this staircase as an open place, it is possible to meet with one another naturally and to observe easily.

MOTIVE

지하벙커가 한 공간에서 외부와 차단되어 지내야하는 유람선, 교도소와 닮았다고 생각되었다. 그것을 바탕으로 우리는 크루즈와 교도소에서의 생활을 착안하여 디자인하게 되었다. 특히 판옵티콘은 많은 사람들을 효과적으로 감시하기 위해 나타난 것으로 교도소에 적용되어 있어 우리의 지하벙커 컨셉과 많은 점이 닮아있었다.

I also thought that the bunker looks like prison or cruise ship that was limited to one space. Based on that, I came up with the idea to design life in cruise and prison. Especially, panopticon has been applied to prisons effectively to monitor many people, and many of them were similar to our concept of the underground bunker.

DESIGN PROCESS

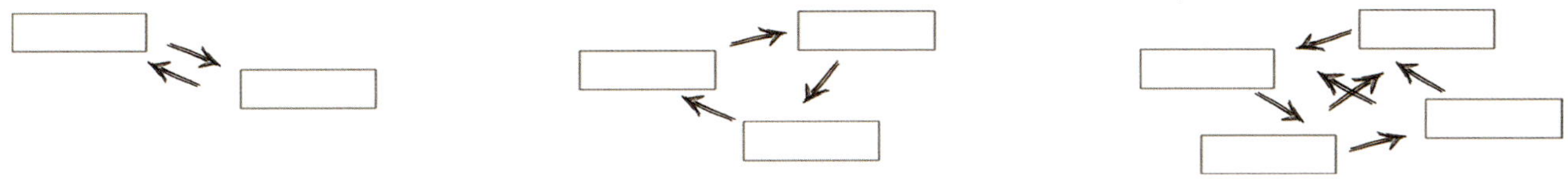

초기 서로 마주하는 방식으로 진행하면서 점차 다시점을 적용하여 공간을 구획했다. 또 높낮이를 다르게 하여 다시점의 폭을 넓혔다.

As we proceeded in a way that confronts each other in the beginning, we gradually applied the points again to divide the space. Also, the height was changed to make th e point more wide.

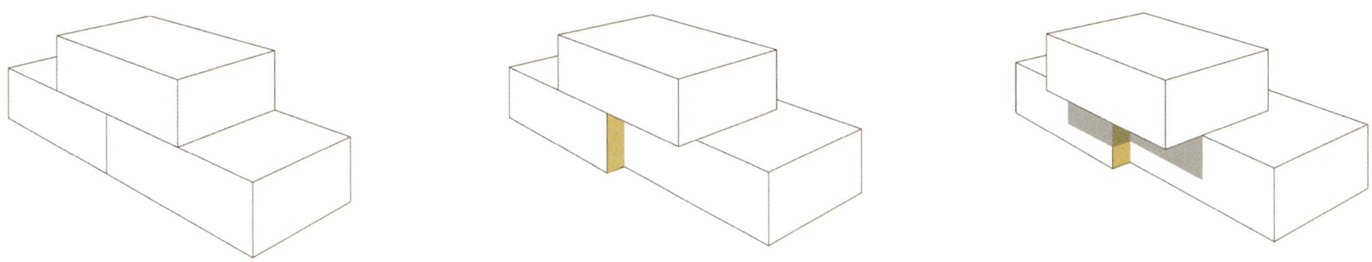

주거공간의 경우 나란히 붙어 있는 형태에서 앞뒤로 뒤죽박죽하게 배치하여 주거공간에서도 계단의 특징을 살렸다. 이로써 위로 배치되는 주거공간의 시야가 넓어지면서 판옵티콘 컨셉 또한 적절하도록 하였다.

As for residential space, we emphasized the key feature of the staircase by rearranging it from side by side way to mixed way. This made the panopticon concept more appropriate as the view of the residential space became wide.

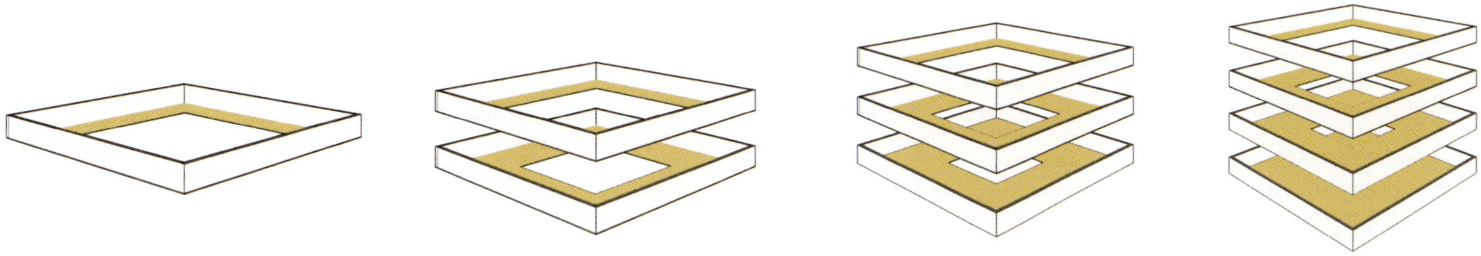

실내에 들어가는 계단뿐만 아니라 건물 자체로도 계단의 특징을 넣었다. 위로 갈수록 디딜 수 있는 면적이 좁아지고, 아래를 내려다볼 수 있도록 하면서 판옵티콘의 특징도 살렸다.

In addition to the stairs entering the interior, the building itself also features stairs. The area that can get closer to the top becomes narrower, and the feature of the pano pticon is saved by making it look down.

1st FLOOR PLAN

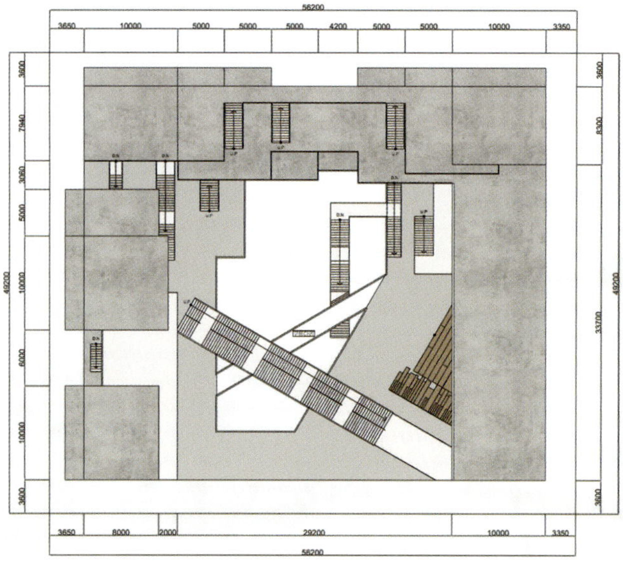

1층은 주거공간과 식당, 강의실, 독서실 등 생활에 필요한 가장 기본적인 공간과 프라이빗을 필요로 하는 공간들을 우선적으로 배치했다.

Firstly, we displayed residence, bistro, classrooms and library, private place and places that we need fundamentally.

B1 FLOOR PLAN

지하에는 도서관과 운동공간, 창고, 공원 등의 공용공간을 배치했다. 지하벙커 안에서 찾아보기 힘든 자기계발과 여가생활을 위한 공간을 둠으로써 삶의질을 높였다.

Public places such as the library, gym, storage, park were displayed in the basement. Moreover, we improved the quality of life by making room for self-improvement and leisure, which is rarely found in underground bunkers.

JONING

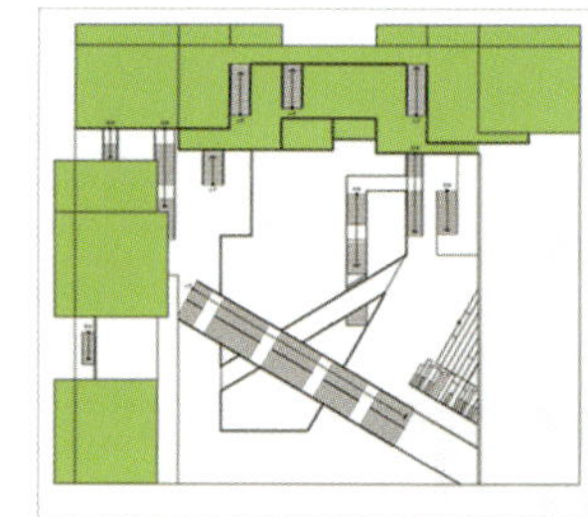

주거공간
Residential space

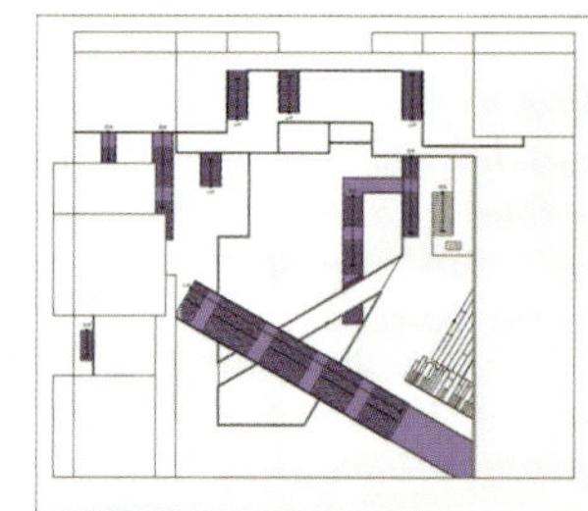

계단공간
Stairwell

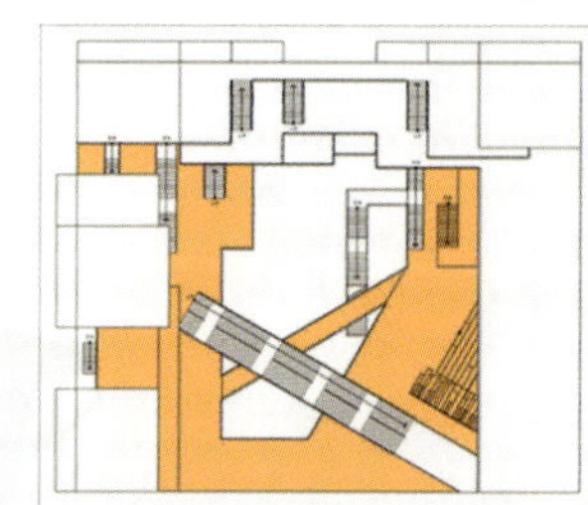

공용공간
Public space

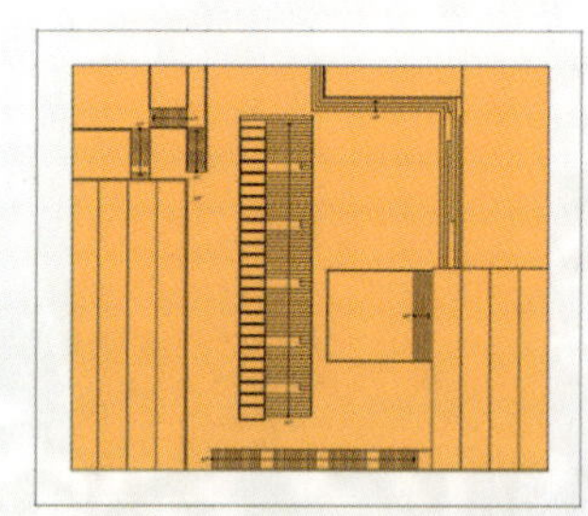

공용공간
Public space

ISOMETRIC

주거공간

인 수에 맞는 침대와 책상하나 화장실 하나로 구성된 주거공간이 6인실 6개, 4인실 24개로 최대 132명까지 수용이 가능하다.

There are 6 rooms for 6 people and 24 rooms for 4 people. It can accommodate up to 132 people. Beds and single desks are included.

식당

계단구조의 3층 식당과 식당에 필요한 물품들이 구비되어 있는 창고가 있다.

There is a three-story restaurant with a staircase structure and a warehouse equipped with items needed for restaurants.

영화관

영화를 보거나 영상을 시청 할 수 있는 공간이다.

Here you can watch movies.

놀이방

아이들을 위한 놀이방이다. 전면이 유리창으로 되어있어 부모님이 아이를 관찰하기에 좋다.

This room is for children. All walls are made of glass, that's why it is good for the parents to observe their kids.

강의실 및 회의실

강의를 하거나 모여서 회의를 할 수 있는 공간이다.
2개의 층으로 이루어져 있다.

This is for lecture and meeting. It consists of 2 floors.

도서관

4층으로 이루어진 계단 형태의 도서관이다.
책과 공부를 할 수 있는 책상이 있다.

It is a library of stairs in layers.
There are books and desks for studying.

공원

산책을 할 수 있는 산책로와 벤치를 두어 공원을 조성했다.

We built a park with walkways and benches for walking.

운동공간

농구와 배드민턴 코트를 두어 운동을 할 수 있다.

Basketball and badminton courts are available for exercise.

창고

기본 생활에 필요한 물품들이 구비되어 있는 창고이다.

It is a warehouse equipped with items necessary for basic living.

PROGRAM

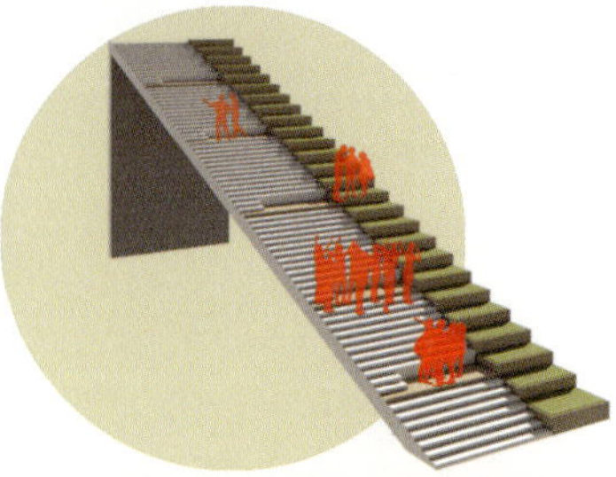

소통
Communication

전체적으로 커다란 오픈공간인 계단이라는 구조물들로 이루어진 지하벙커에서 사람들과의 만남은 필수적이다. 이는 계단이 만남과 소통을 자연스럽게 유도하는 하나의 장치로도 사용 될 수 있음을 보여준다. 또한 커다란 계단공간과 같은 공간에서 모여서 논의를 할 수 있으며 토론과 소통의 장을 만들 수 있다. 계단 위에서 사람들의 위치를 파악하기도 용이하다.

Meeting people is essential in an underground bunker with a structure called the staircase, which is entirely a huge. This shows that the staircase can also be used as a material that naturally leads to meetings and communication. They can also gather together in the same space as a large staircase, and create a forum for discussion and communication. It is also easy to locate people on the stairs.

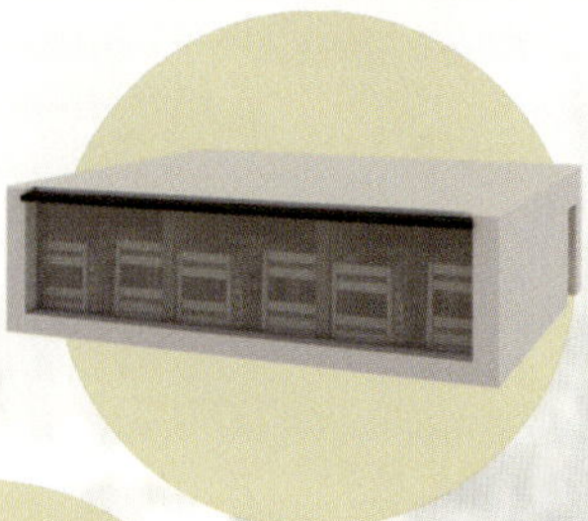

개인공간
Private Space

판옵티콘이라는 개념과 계단의 특성이 합쳐지면 원활한 관찰을 위해 공간이 공개적일 수밖에 없다.그러나 삶의 가장 기본이 되는 주거공간에서의 프라이빗함을 보장했다. 내부에서 바깥을 관찰 할 수는 있지만 바깥에서 내부를 관찰 할 수는 없도록 특수유리를 사용해 프라이빗을 보장한다. 또한 주거공간을 가족단위로 배치해 모르는 외부인에게 사생활을 노출해야하는상황을 막는 동시에 심리적 안정감을 줄 수 있다.

The combination of the concept of the panopticon and the characteristics of the staircase opens the space for smooth observation. But it surely guarantees privacy in the most basic living space. You can even observe theoutside from the inside but the special glass ensure the inside is not visible from the outside. In addition, the allocation of residential space on a family level could provide a sense of security and safety while avoiding the need to expose privacy to strangers.

다목적
Multi-Purpose

계단을 층의 연결수단, 이동수단으로만 이용하는 것이 아니라 계단을 공간으로 보고 의미를 부여해 다목적성을띄게 한다. 예를 들면 계단이 도서관으로 이용 될 수도 있고 영화관, 강의실, 모임장소 등으로 이용 될 수도 있다.이를 통해 다목적성을 갖게되고 계단의 의미와 필요성을 한 단계 발전 시킬 수있다.

Instead of using the stairs as a means of connecting the floors only, the stairs are regarded as space, giving the meaning to a multipurpose as well. For example, stairs can be used as a library or movie theater, lecture room, or meeting place. This provides versatility and necessity to the meaning of the staircase.

다시점
Multi-View

계단을 정형화 된 형식으로 설치하는 것이 아니라 공간과 공간 사이를 이어주도록 자연스럽게 설치하여 다시점을 강조해 공간의 단조로움을 해소시켰다.

Stairs were installed naturally to connect space to space rather than to form a formal installation, hence, we reduced the monotony of the space by highlighting the multi view points.

여러 개의 계단이 겹치고 가로지르며 여러 개의 시점을 만들어 낸다. 수 많은 계단에서 바라본 시선이 각각의 계단마다 다르고 위치에 따라 변화한다. 그에 따라 비슷해 보이면서도 다른 분위기를 연출한다.

Multiple stairs overlap or intersect to create multiple views. The visibility of the various stages varies from one stair to another, depending on location. It looks similar but creates a different atmosphere.

1층과 2층을 연결해주는 거대한 계단은 소통의 역할을 한다.
A huge staircase connecting the first and second floors serves as a means of communication

계단이 주거공간이 되기도 하고 영화관으로 사용되는 등 다양한 목적성을 갖는다.
The stairs have the various multi-purpose. It can be a residential space and movie theater as well.

계단의 형태를 한 강의실이다. 막힌 공간이지만 유리로 되어있어 개방감을 주고 관찰하기에도 좋다.

It is a lecture room in the shape of a staircase. Even though it is a blocked space, but since it is made of glass, it is good to give an openness assessment and to observe.

운동공간 Exercise space

농구장과 배드민턴장 클라이밍 등을 두어 운동을 즐길 수 있는 공간을 만들었다. 지루 할 수 있는 지하벙커 생활에 활기를 불어넣을 수 있다.

Basketball courts and badminton courts are installed in this space to enjoy the exercise

도서관 및 지하공원 Library and Basement park

지하 층에 위치한 도서관과 공원이다. 도서관은 4개의 층으로 이루어진 계단 공간으로 책과 더불어 공부를 할 수 있는 책상이 마련되어 있다. 지하공원에는 산책로와 벤치를 두어 삭막한 지하공간에 생기를 불어넣었다.

A library and park located on the basement floor. The library is a four-story building with desks and books. The underground park has its walking trails and benches to refresh.

지하계단 Basement stairs

지하 공용공간의 4개층을 1층으로 이어주는 계단이다. 계단이 겹쳐져 있는 모습에서 반복성을 느낄 수 있다.

It connects the four floors of the public spaces under the ground to the first floor. You can feel repetition from the overlapped of stairs.

중앙계단과 연설대 Central Stairwells and Speeches

1층과 지하를 이어주는 이 거대한 계단은 지하벙커안의 모든 사람들을 수용할 수 있어 모두가 모여 논의를 하거나 토론과 소통을 할 때 유용해 일종의 광장과 같은 역할을 한다고 볼 수 있다. 또한 연설대는 계단의 중앙에 위치해 소식을 알리는 용도로 사용 할 수 있다.

This huge stairway that connects the first floor with the underground can accommodate everyone in the underground bunker and is useful for gathering, and to communicate with everyone. It can be regarded as a sort of plaza. Furthermore, the podium is located at the center of the stairway. So we can utilize it in many ways.

강의실 및 회의실 Classroom and Meeting Room

2개의 층으로 이루어져 있으며 커다란 계단의 형태를 하고있다. 강의실 및 회의실로 사용이 가능하다. 강의실이 바깥쪽 벽이 유리로 되어있어 막힌공간이지만개방되어있는 느낌을 준다.

It is made up of two floors and is shaped like a large staircase. We are allowed to use those as classroom and conference room. Since the outer side of the lecture room is made of glass, it gives the feeling of openness though it is blocked.

식당 Restaurant

3층의 식당과 1층의 창고로 이루어져 있으며 계단의 형태를 하고 있다. 각 가구별로 요리를 해먹을 수 있는 수방과 식탁이 마련되어있다. 판옵티콘적 요소를 반영해 구성한 계단의 형태는 가장 윗층의 식당에서 가장아랫층의 식당을 관찰하기에도 무리가 없다.

It consists of a restaurant on the third floor and a warehouse on the first floor. It is in the shape of a staircase. Each kitchen has a dining table and kitchenware for each household. The shape of the staircase, which reflects the panopticon allow you to observe the restaurant from the lowest floor to the highest floor.

놀이방 Playroom

아이들을 위한 놀이방이다. 전면이 유리로 돼 있어 놀이방 안에서 밖을 관찰하거나 바깥에서 놀이방 안을 관찰하기에 무리가 없다.

Here is the playground for kids. All walls are made of glass, so it is easy to observe them from inside to outside and from outside to inside.

영화관 Cinema

영화를 보거나 영상을 시청 할 수 있는 공간이다. 지하벙커 안에서는 찾아보기 힘든 시설로, 여가생활을 즐길 수 있는 요소를 마련 함으로써 삶의 질을 높였다.

It is a space where you can watch movies. By making this in the underground bunkers, we tried to improve the quality of life for enjoying leisure activities.

Senepi : Sense of Epidemic

전염병은 감염 중에서도 그 전염성이 강하여
소수의 병원체로도 쉽게 감염되고
많은 사람들에게 쉽게 옮아가는 질병을 말한다.
문명이 발달할수록 전염병도 새로이 발견되고
변형되는 모습을 보여병의 완전한 제거는 어려운 상황이다.
따라서 우리는 전염병에 대한 공포를 인식시키고
난 안 걸리겠지 "라는 안전불감증에 대한 의식을 개선시키고자 한다.
우리는 전염병이라는 질병을
스토리텔링하기 위해 총 5개의 공간을 구성하였다.
이 공간들은 빛과 어둠, 소리, 향기들로 채워져 있으며
이러한 매개체들로 감각을 자극하여
시민들이 감각으로 하여금
전염병의 확산을 지각할 수 있도록 했다.

First of all, an infectious disease is a disease that spreads
very easily to people.
It is easily infected by a small number of pathogens.
And It is difficult to eliminate the disease thoroughly
because it keeps changing as civilization develops.
Therefore, we intend to raise our awareness of the dangers of
infectious diseases and to raise consciousness that "we are safe
from all dangers. "
Thus, we have formed
five places to explain the disease of epidemics.
These spaces are filled with light, darkness, sound, and smell, and
these mediators stimulate the senses,
allowing sensory organs to recognize the dispersion of the disease.

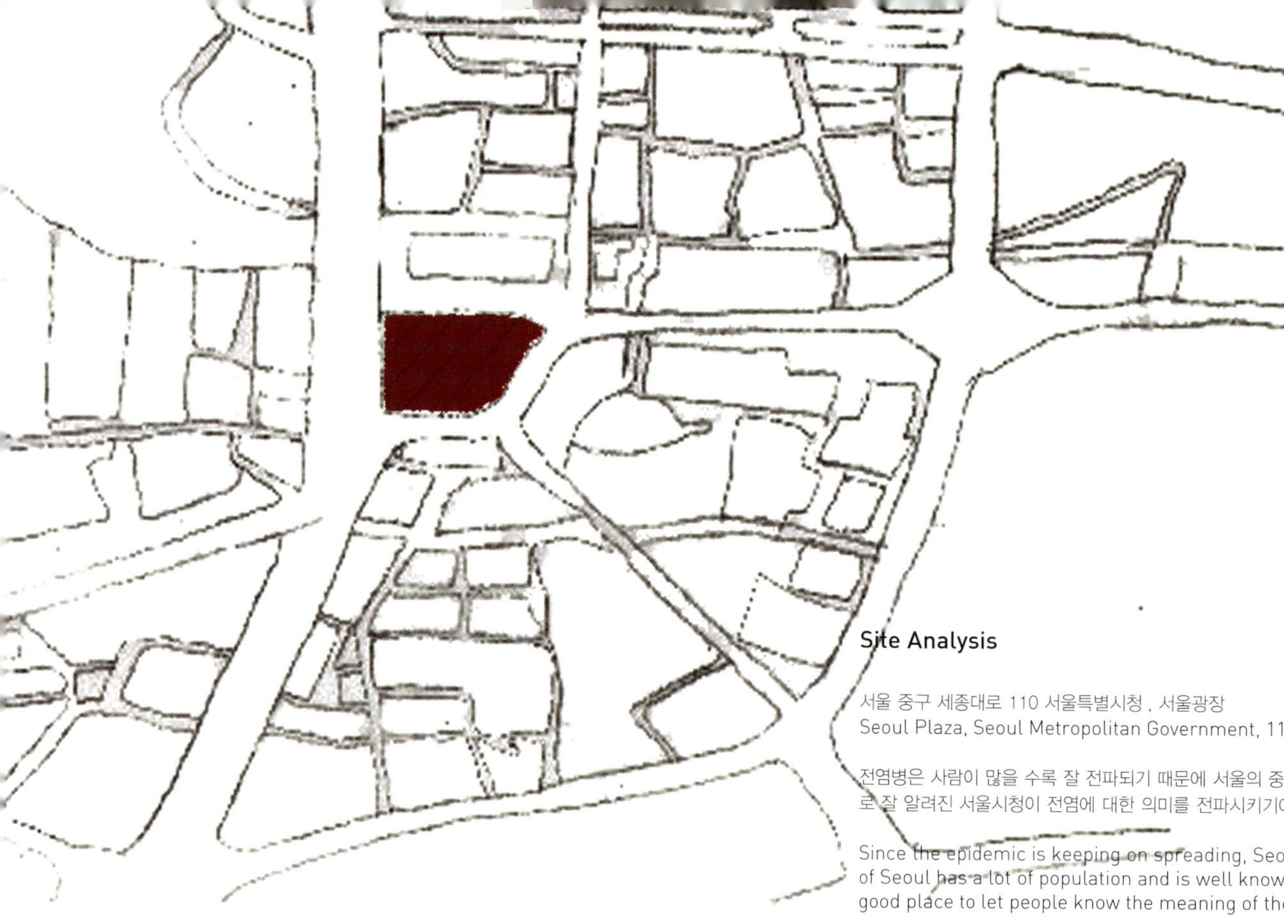

Site Analysis

서울 중구 세종대로 110 서울특별시청 , 서울광장
Seoul Plaza, Seoul Metropolitan Government, 110, Sejong-daero, Jung-gu, Seoul

전염병은 사람이 많을 수록 잘 전파되기 때문에 서울의 중심지이자 유동인구가 많고 위치적으로 잘 알려진 서울시청이 전염에 대한 의미를 전파시키기에 좋은 장소라고 생각했다.

Since the epidemic is keeping on spreading, Seoul City Hall, which is the center of Seoul has a lot of population and is well known locally, we thought that it is a good place to let people know the meaning of the infection.

Design Concept

지구에 바이러스 물체가 떨어지고 지구와는 어울리지 않는 모양을 가진 이 물체는 이질적이며 괴리감마저 느끼게한다.
하늘로 치솟을 것 같은 벽체와 내부가 투명하게 내비치는 돔계단, 위태로워 보이는 계단은 위험해보이고 불안하다.
우리는 이러한 건물의 특성을 살려 전염에 대한 공포와 그 속에서 체험할 수 있는 공간의 흥미를 불러 일으키고자 한다.

This object, which has a viral object falling on the earth and has a shape that does not match to the earth, makes us even feel eerie.
The interior is transparent and those stairways look dangerous and unstable.
We would like to use the characteristics of these buildings to inspire fear of infectious diseases and make people interested in what they can experience there.

Mass Process

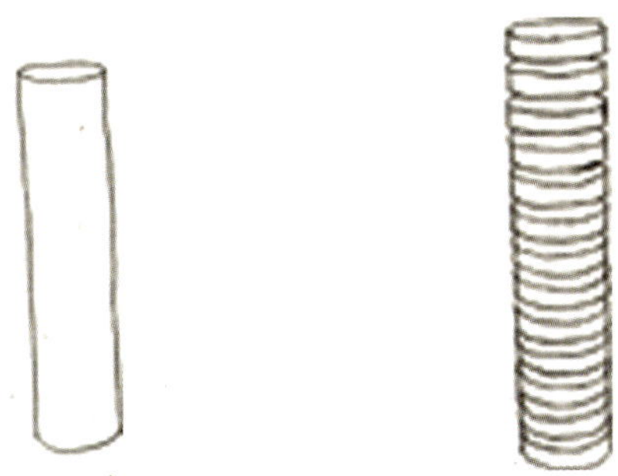

공간삽입하여 단면에 다공성 부여
Porosity design with space insertion

면적의 확산
Diffusion of area

중심축에 다공성 부여
Setting porosity design at the center axis

바이러스 덩어리
A Virus Mass

내적 분열
Internal division

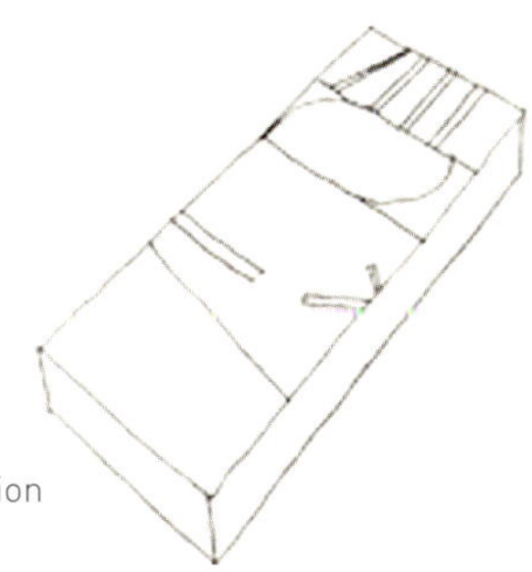

외적 분열
External division

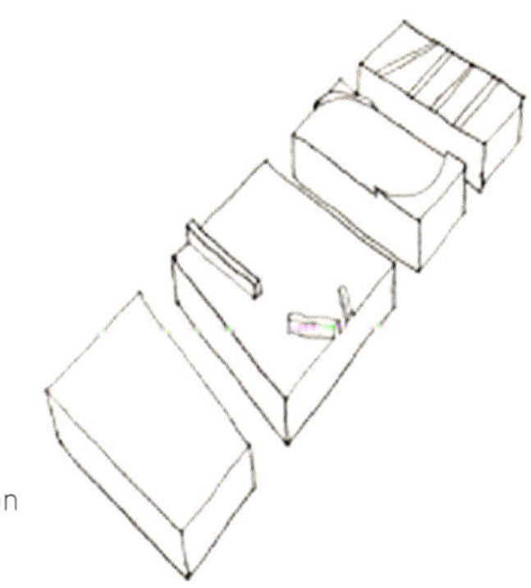

Furniture Process

SENEPI의 유기적인 구성에 방해되지 않고 유연한 시퀀스를 진행가지도록 하며 편안한 휴식을 동반할 수 있는 가구

Furniture that has a flexible sequence and that can also be suited with comfortable rest without disturbing the organic composition of SENEPI.

Plan

Elevation & Section

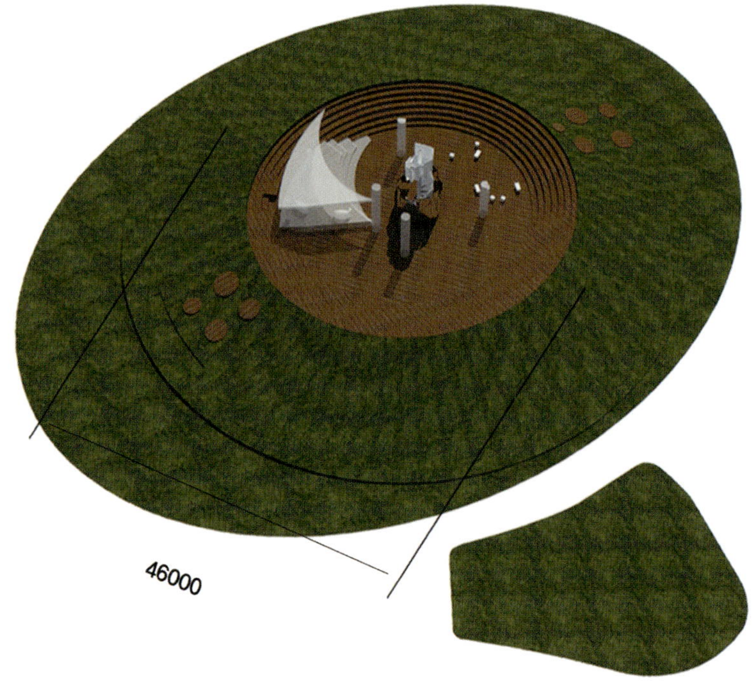

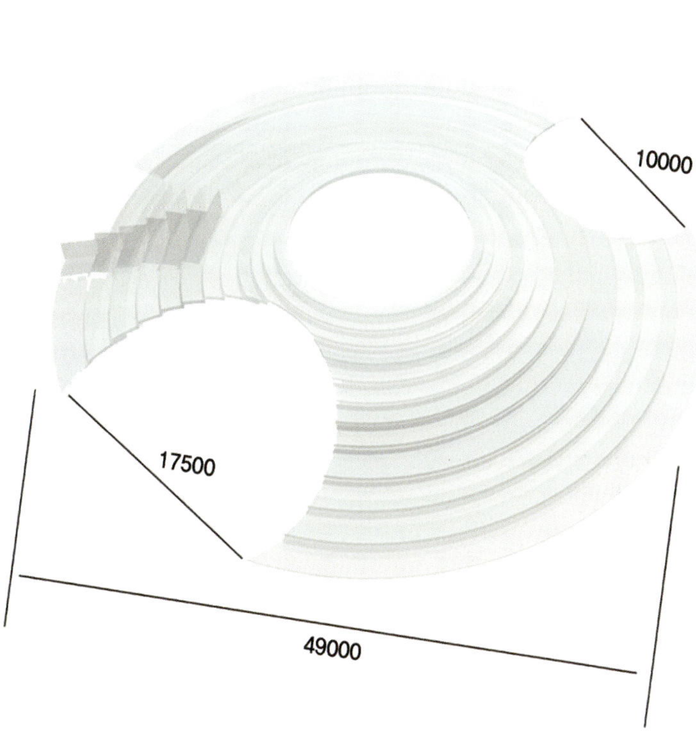
10000
17500
49000

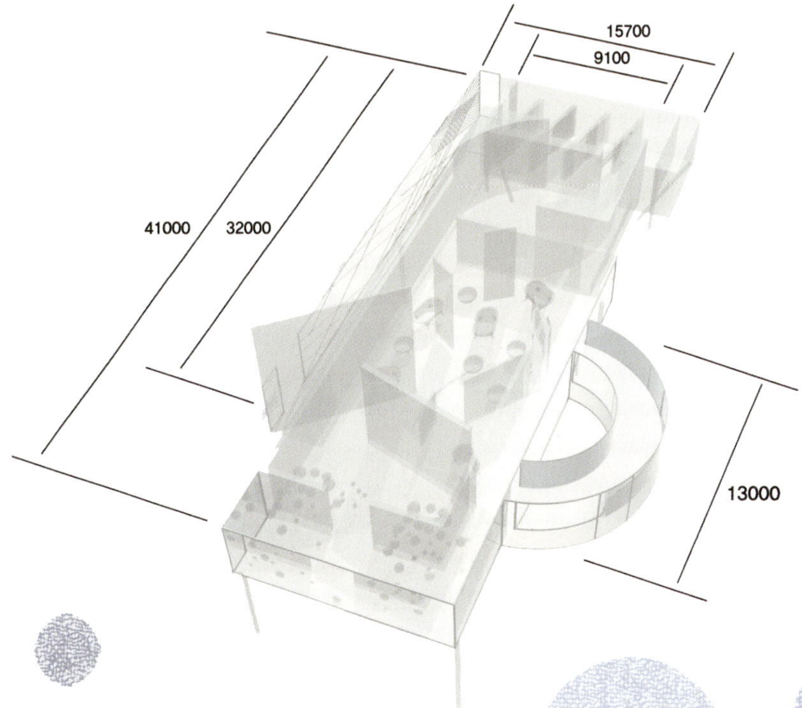
15700
9100
41000
32000
13000

확산의 감각화 The Sensation of Dispersion

전염병의 특징인 확산에 집중해 확산에 대한 공포를 감각으로 지각하게 함으로 전염병이란 누구에게나 걸릴 수 있다는 것을 신비로우면서도 다각적인 면으로 나타냈다.

We have shown it in a mysterious and multifaceted way, letting people pay attention to the characteristic of an epidemic and we have shown that infectious diseases can be carried away by anybody mysteriously and in a multi-culture way.

다공 Porous Design

막힌듯 뚫려있는 다공성의 이중성이라는 특징을 전염병에 투영하여 불안이라는 감정과 신비로운 느낌을 표현했다. 이로서 관람객이 직접 공간을 체험하며 공간에 풀어낸 감각들을 흡수하는 과정에서 불안과 신비로움이라는 이중성을 느낄 수 있다.

The characteristics of porous duality are projected into infectious diseases to express the feeling of anxiety and mysticism. As a result, visitors can feel the duality of anxiety and mystery in the process of experiencing space directly and understand the senses as they get into the place.

Entrance

Entrance 1 _ A Low Grass Hill
둔덕을 타고 돔 안으로 들어가는 입구
An Entrance into the dome on the grass hill.

Entrance 2 _ Dome Stairs
돔계단을 이용해 2층으로 들어가는 입구
Entrance to the 2nd floor using these stairs.

On the ground floor

나선계단 _ The Circular Staircase
앞만 보고 올라가도 앞을 정확히 알 수 없는 나선계단. 이것은 이전과는 다른 새로운 전염병을 맞이할 수 있음을 의미한다.

people cannot see exactly what is ahead as they go up. This informs us that we might be facing a new kind of epidemic.

실문 & 실벽 _ Lined door and wall
실벽은 그 다음의 공간에 대한 금증을 유발한다.

The walls of the room cause curiosity about the next place

계단의 확산 _ Stairs Spread
1층을 둘러싸고 있는 원형의 계단은 일부의 공간에서 확장되어 실문과 만난다. 이것은 마치 평온한 일상에서 전염병이 퍼져 막연했던 두려움과 마주하는것과 같다.

The Circular staircase, which surrounds the first floor expands in some space and connects directly to the door. This can make people feel that they are facing an unimaginable fear that spreads from an epidemics of daily life.

거울 _ Mirror
거울은 가상의 같은 공간을 하나 더 만들어 공간을 확장하여 보이게 한다. 병원체의 잠복기와 같다.
The mirror creates one more virtual place, which can be seen by expanding it. This place gives us the feeling that visitors are in an incubation period of a pathogen.

On the First floor

바이러스의 잠복기 _ The incubation period of a virus

바이러스의 잠복기를 표현한 공간으로 바이러스를 공으로 묘사해 철창의 안과 밖에 배치하여 시각화하였다.

We visualized the virus by portraying it through balls on the ceiling. Therefore it would be described the incubation period of the virus in and out of wire entanglements.

바이러스의 잠복기 _ The incubation period of a virus

이제 관람객이 건물을 침투한 바이러스가 된다. 청결을 표현하면서 벽을 아무렇게나 세워놓은 공간으로 아무리 위생 행동을 취해도 바이러스들은 남겨져 다음 공간으로 넘어간다.

Imagine that visitors become viruses that penetrate buildings. As the room stands for cleanliness, no matter how hygienic the action is, the virus is left in the next room.

바이러스의 전염성 _ Dark Spread

마지막 공간은 처음 밝았던 공간에서 문을 열어 갈수록 어두워져 간다. 전염병으로 인한 죽음을 인식시키게 하고 밖으로 나가는 순간 관람객들을 다른 사람들에게 전시를 전염시키는 바이러스 역할을 맡게 한다.

The last place gets darkeras you open the door from the place where it was first opened. This place allows the viewer to recognize the death from an epidemic and the viewer plays a role like a virus that spreads the exhibition to others as they leave the place.

바이러스의 확산기 _ The Diffusion of Virus

다음 공간으로 넘어간 관람객들은 빔프로젝터로 꾸려진 어두운 공간으로 가게 되는데 바이러스들이 떠다니는 공간을 체험함으로써 바이러스의 확산과 전염성을 느끼고 누구에게나 병이 걸릴 수 있다는 메시지를 경각할 수 있다.

Visitors who pass through room go into a dark space, which is set up by beam projectors. By looking around the room, where the virus floats, we can feel the spread of viruses and epidemics. Through this place, we want to give visitors the awareness that anyone can be ill.

Exit

전시의 설명을 듣지 않았을 때 전염병에 대한 전시라는 것을 모르는 관람객들도 있을 것이다. 우리는 전시의 내용을 강요하지 않을 것이다. 관람객들이 전시로 인해 스스로 가지는 감정, 느낌들을 전유하고 생각할 수 있는 시퀀스를 나타내 전시의 키워드였던 다공으로 표현하면서 전시를 마무리했다.

Some viewers may not know about the purpose of this exhibition when they do not understand the explanation clearly. We are not forcing them to understand the contents of the exhibition. We concluded with a sequence that visitors are allowed to think about their feelingsof the exhibition that we tried to express virus through differentkinds of rooms.Lastly, we ended up our exhibition by reminding our key concept, which is porosity.

염은진 I Yeom Eunjin, 서정인 I Seo Jeongin

RESCUE DOMINO
:재난건축 7원칙
7 principles of construction about a disaster

이번 주제는 재난, 또는 재난을 위한 shelter이다. 다양한 형태로 나타나는 재난을 위한 대피소의 개념을 공간에 풀어내고자 위의 주제를 이용해 재난 중 '조난' 을 선택했다. 산에서 일어나는 크고 작은 사건 사고들을 대비하고 사용할 수 있으며 기존 산악 대피소와는 차별화된 shelter를 제시한다.

Since the concept of this international exchange exhibition is about a disaster or shelter for disaster, we chose 'distress' in order to explain the concept of shelter for various disasters in place. We have prepared for accidents whether according to large and small incidents, which can happen in mountains. And we are going to present a shelter that is different from the normal one.

BACKGROUND

우리나라는 70%가 산이라는 말이 있듯, 산을 찾는 등산객도 점점 증가하는 추세이다. 그러나 현재 지리산이나 설악산 국립공원에 설치된 대피소들은 예약을 하고 가야 하며 상인들과 북적이는 사람들로 가득해 산장과 같은 역할만 하고 있으며, 정작 대피소의 구실을 못하고 있는 상황이다. 진정한 의미의 조난 사고를 방지하는 등반가와 캠퍼들을 위한 임시 산악 대피소를 진행하려고 하며, 우리나라뿐만이 아닌 어느 산이어도 설치가 가능한 디자인이자 대피소 역할은 물론 자가발전 시설을 갖춘 첨단 대피소이다. 계획 대지는 우리나라에서 세 번째로 높은 산이며 세계적인 보존지역이자 관광지인 '설악산'으로 선정하였다. 설악산의 있는 기존 대피소들을 중심으로 마치 위성도시처럼 곳곳에 설계한 대피소들이 놓인다면 조난의 피해가 감소할뿐더러 자연과 인간이 교감할 수 있는 공간으로 적격이다.

As 70% of Korea is covered by mountains, the rate of mountain climbers is kept on rising as well. However, most of the shelter installed at Jiri Mountain and Seorak National Park is packed with merchants and crowds and do not function as shelters. We intend to set up temporary mountain shelters for climbers and campers who really need the real shelter. This shelter is made up for self-engineering, self-engineering shelter, which can be installed not just in our country, but for other countries as well. The location that we planned for is the third highest mountain in the nation - Seoraak Mountain. This is because Seolak Mountain is designated as the world's best-preserved area and it is one of the famous tourist sites in South Korea. Moreover, if the shelter would be built around other shelters in Seorak Mountain, the damage from the wreck would decrease and it will be the best place for people and nature to interact with each other.

MOTIVE

르코르뷔지에는 사보에 주택을 설계하기 전 1차 세계대전 이후 황폐화된 도시를 어떻게 하면 사람들에게 더 빠르고, 많은 집을 지어 줄 수 있을까 고민하였고 그 결과 새로운 건축 방식인 도미노 시스템이 만들어졌다.

Le Corbusier wondered how he could build peoples' houses faster and more easily the devastation of the city after World War I. He came up the result to build it with a new architecture method of domino.

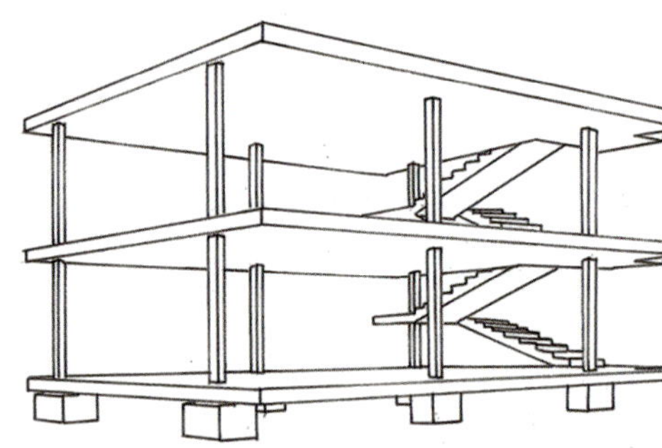

르코르뷔지에의 근대 건축 5원칙에는 수평 창, 자유로운 평면, 자유로운 입면, 옥상 정원, 필로티가 있으며 조난 상황에 맞도록 기존 5원칙을 토대로 총 7가지의 원칙을 공간에 적용하고자 한다.

There are five principles of modern architecture include horizontal windows, free planes, free facades, roof gardens and piloti. We would like to apply 7 principles to the space based on 5 principles according to different types of situations.

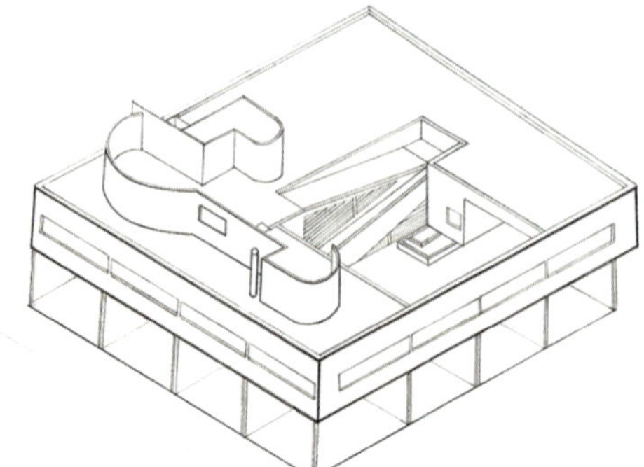

DESIGN PROCESS

주어진 대지에 맞게 수직적 부재를 꽂는다. 수직적 부재에 의해 생겨난 공간에 프레임을 연결하여 매스의 형태를 만들고 프레임 사이에 나무 벽을 조립한다. 공간과 공간을 이어주는 계단과 테라스가 형성되며 자가발전 시스템 풍력터빈이 설치된다.

First of all, insert a vertical member to match the given site. Then, connect the frame to the space formed by the vertical member to form a mass and assemble the wooden wall between the frames. Stairs and terraces that connect space to space will be formed and wind power turbines of a self-generating system will be installed.

CONCEPT : 재난건축 7원칙

1. 필로티 Piloty – 수직적 부재 Vertical absence
: 근대 건축 5원칙에서 필로티는 주차장으로 이용되지만 재난 건축 7원칙에서는 산의 능선에 맞게 수직적 부재를 세워 공간을 형성하게 한다.

Pilotis are used as parking lots in modern architecture 5 principles. In 7 principles, disaster building, vertical materials will be set to form a place according to the mountain ridges.

2. 철제 프레임 Iron frame – 뼈대 skeleton
: 수직적 부재에 의해 형성된 공간에 철제 프레임을 조립하여 뼈대를 만든다. 이로 하여금 빠르고 견고하게 공간이 만들어진다.

The frame is constructed by assembling steel frames into spaces formed by vertical materials. This makes the space faster and stronger.

3. 에너지 – 풍력터빈
: 풍력터빈을 설치하여 산속에서 쉽고, 강력하게 많은 양을 얻을 수 있는 바람 에너지를 통해 전기 에너지를 저장하고 난방과 조명 그리고 온수를 공급해 줄 수 있도록 한다.

Next, we installed a wind turbine to save electricity. It provides heating, lighting, and hot water through wind energy that is easily obtainable and available in the mountains.

4. 응급시설 Emergency facility – 인큐베이터 Incubator
: 조난 상황에서 부상자를 위한 응급 공간이다. 인큐베이터 안에서 부상자가 심리적 안정감을 느끼며 연결된 통로를 이용해 헬기로 신속한 이송을 할 수 있도록 설계한 시스템이다.

We designed the system to enable injured people to move quickly by helicopter. The path that we installed in case of emergency, the injured could feel comfortable psychologically.

5. 자유로운 단면 Free section – 1층과 2층 1f 2f
: 1층– 실내의 공간 구획을 사용자에게 맡기기 위해 불필요한 벽이 아닌 단 차이로 공간 분할을 했다.
2층– 하중을 받지 않는 내벽의 위치를 자유롭게 처리하여 높은 층고를 활용해 복층 구조로 만들었다. 이로 인해 남는 공간 없이 최대한의 공간 활용이 가능하다.

This place is divided into two parts. We only used a column to separate those. The wall inside is made of the multi-layer structure by utilizing layer to layer in the building. As a result, maximum space will be utilized without the need for extra space.

6. 옥상정원 roof garden – 전망대 Observatory
: 근대 건축 5원칙 중 하나인 옥상정원은 전망대로 활용이 가능한 공간과 휴식 공간으로 구성된다. 재난 상황이긴 하지만 shelter의 개념으로 안식과 휴식을 취할 수 있도록 한다.

A Roof garden consists of a space for observatory and resting space according to the 5 principles of modern architecture. You can even take a rest in the area of a refuge.

7. 모듈러 Modular – 가구 furniture
: 2층 침실의 작은 평수와 높은 층고의 특징을 살려 블록처럼 쌓아올린 매스의 반복을 통해 생겨난 공간으로 최대한의 인원을 수용하도록 가구디자인을 하였다. 작지만 견고하고 아늑한 느낌을 준다.

The furniture was designed to accommodate the maximum number of people in the space created by the repetition of the mass that piled up like a block shape. Small sizes of beds are installed. It is small but makes us cozy..

ISOMETRIC

휴식 Relaxation
층마다 산이라는 장소적 특징을 살려 사람과 자연이 교감할 수 있도록 최대한의 느낌을 살리기 위해 테라스를 곳곳에 두었고 옥상정원을 만들어 전망대로 활용이 가능하다.

The important thing is that we use terraces all over the floor, allowing people and nature to interact with each other. Moreover, this roop garden can be Observatory as well.

인큐베이터 The Incubator Building
응급 공간에는 인큐베이터 시스템이 설치되어 있다. 부상자를 실은 인큐베이터는 1층부터 지붕까지 연결된 통로로 이동이 가능하여 헬기가 신속히 이송할 수 있도록 설계한 시스템이다.

The emergency room has an incubator system. A wounded person can move from the ground floor to the roof, the injured could be transported quickly by a helicopter.

공용공간 Public space
공용 공간은 고정된 주방 가구를 중심으로 둘러 싸인 형태이다. 다 같이 모여 공용 공간으로도 활용이 가능하고 부엌의 기능도 있으므로 사용자의 요구에 따라 자유롭게 공간활용이 가능하다.

The public space is surrounded by kitchen furniture. It can also be used as a public space and has a kitchen as well. Therefore we can use it freely according to the needs of the user.

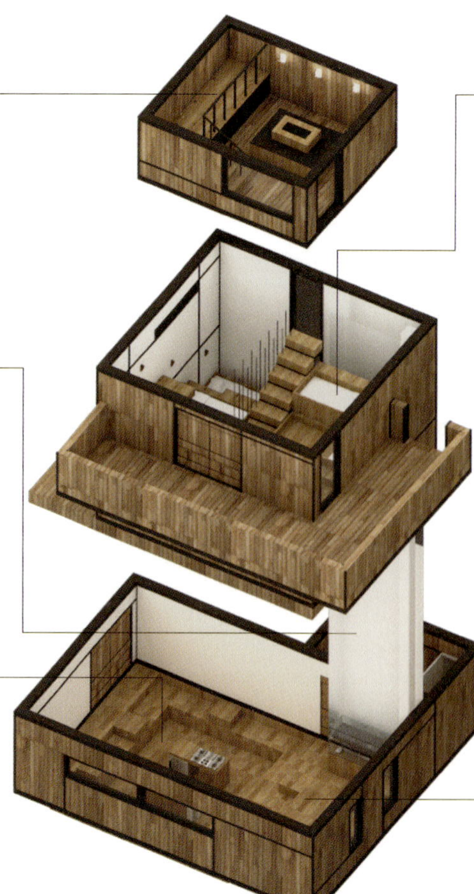

침실 Bed room
사용자가 편히 쉴 수 있는 공간인 침실이다. 총 7개의 침대 가구가 놓여있으며 7~10인까지 수용이 가능하다. 침실의 내부 계단으로 올라가게 되면 옥상 정원과 테라스로 나갈 수 있다.

It is a bedroom that accommodates 7 people. 7 beds are available and can be used by 7 to 10 people. You can go directly to the rooftop garden by using the staircase inside the bedroom.

응급 공간 Emergency space
공용 공간 옆에는 디자인 프로세스 중 하나인 응급 공간이 형성됐다. 대피소의 목적이자 가장 핵심적인 공간인 응급 공간은 1층에 배치되어 사용자로 하여금 신속하게 치료와 이송을 이루어지게 한다.

Next, to the public space, an emergency space is made. The emergency space is located on the first floor, allowing the user to be transported quickly.

ISOMETRIC

[1F: Public Space, Emergency Space, Bath room / 2F: Bed Room, Terras/ Roof Garden]

PERSPECTIVE01 [1F]

1층은 공용 공간과 응급 공간으로 이루어져 있으며 근대 건축 5원칙의 자유로운 평면이 잘 나타나도록 벽이 아닌 단 차이를 두어 공간을 분할했다. 부엌의 기능을 갖춘 공용 공간에는 사용자에 따라 나무 가구가 이동이 가능하여 자유롭게 앉기도 하고 식사는 물론 다양한 공간 활용이 가능하도록 디자인하였다. 응급 공간에는 등산 도중 부상자를 위한 공간으로 인큐베이터를 공간 안에서 치료할 수 있는 공간이자 신속히 이송할 수 있도록 설계한 시스템이다.

The first floor consists of public spaces and emergency spaces, and spaces were divided into two separate sections by columns. So that the five principles of modern architecture can be seen well. After that, we designed the place, where it has a kitchen, the wooden furniture can be moved to be used freely by sitting and having meals there.
Emergency space was designed for the injured people to be treated while climbing.

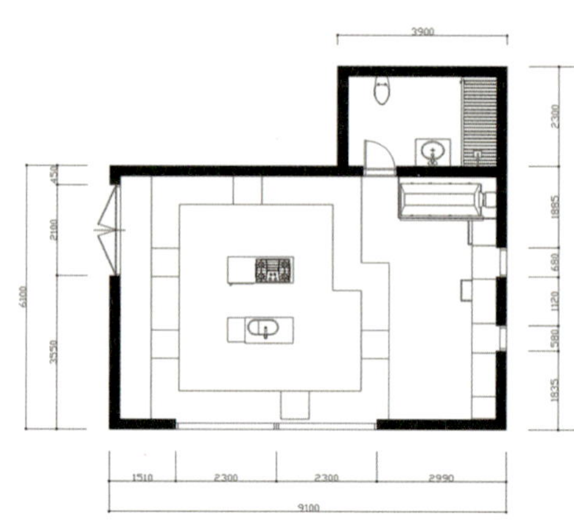

PERSPECTIVE02 [2F]

1층에서 2층은 외부 계단으로 올라가게 된다. 2층은 침실 공간으로 작은 평수와 높은 층고의 특징을 살려 모듈 시스템으로 디자인했다. 벽이 없는 자유로운 평면에 맞추어 계단 난간을 길게 디자인하여 가벽의 느낌을 주고 침실로서 프라이버시와 공간 분할을 했다.

We use the stairs outside to go up from the first floor to the second. The second floor was designed as a modular system by showing key features of small space and high floor like a bedroom place. The stairs are designed long with free plane without building any walls to show the division of the bedroom and the private space.

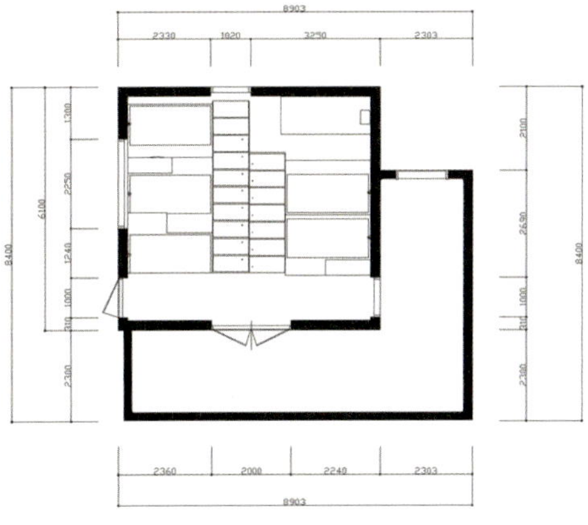

PERSPECTIVE03 [ROOF GARDEN]

근대 건축 5원칙 중 하나인 옥상 정원으로 가기 위해서는 2층 테라스로 나와 외부 계단을 통해 올라갈 수 있다.

옥상 정원은 전망대와 휴식할 수 있는 공간으로 구성되었는데, 자연 느낌을 살리기 위해 나무와 유리로 마감하였으며 자연경관을 볼 수 있게끔 한다. 재난 상황이긴 하지만 shelter의 개념으로 안식과 휴식을 취할 수 있도록 한다.

you can go up the stairs to the second-floor terrace to go to the rooftop garden, which is one of the five principles of modern architecture.

The rooftop garden consists of an observation deck and a rest area. We finalized with wood and glass to create a feeling of nature and let people see the view of nature. Though in a disaster situation, the Shelter provides people to rest.

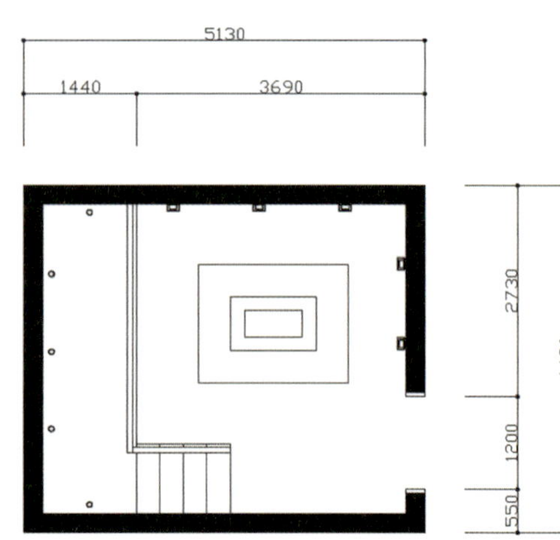

PERSPECTIVE04 [TERRACE]

1층에서 2층을 연결하는 곳에 테라스를 두었으며, 2층의 높은 층고를 이용하여 옥상정원과 연결하는 테라스 공간을 두어 전망대로도 활용이 가능하다. 이를 통해 자연과 사람이 교감할 수 있다. 풍력터빈이 설치된 지붕면에는 경사를 주어 눈이나 비가 자연적으로 흘러가도록 했으며, 반대쪽 지붕면에는 인큐베이터가 헬기로 이송을 위해 올라오는 공간이다.

The terrace is located on the first floor to the second floor, and a terrace, which is connecting to the rooftop garden can also be used as an observation area using the high floor height. It enables nature to interact with. The roof surfaces on which the wind turbines are installed are inclined to let in snow or rain naturally, and on the other side of the roof surface, an incubator can be transported by helicopter.

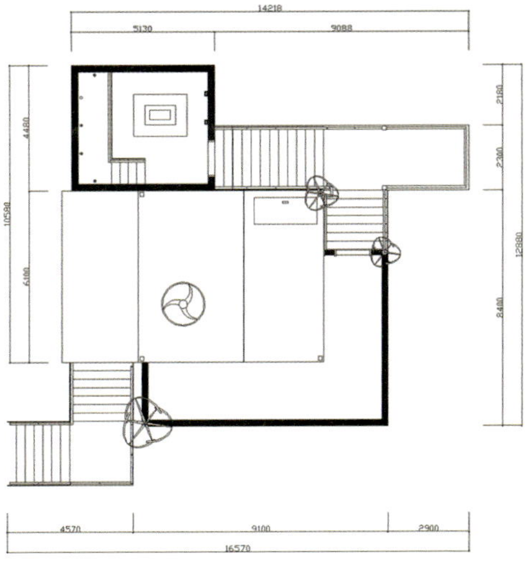

GET A DISASTER ; 재난을 딛다

강한 파괴력을 가진 재해로서의 지진.
더 이상 한반도는 지진 안전지대가 아니다. 평소에는
체험관으로 익숙한 건물이 재난 시 대피소로 사용되는
체험용 지진 대피소를 구상해 보았다.

The most powerful disaster in the world - an earthquake. South Korea is no longer a safe country from an earthquake. We have planned to build a familiar building that can be used as a shelter in the event of a disaster for the experience.

함현석 I Ham Hyeonseok, 김다솜 I Kim Dasom

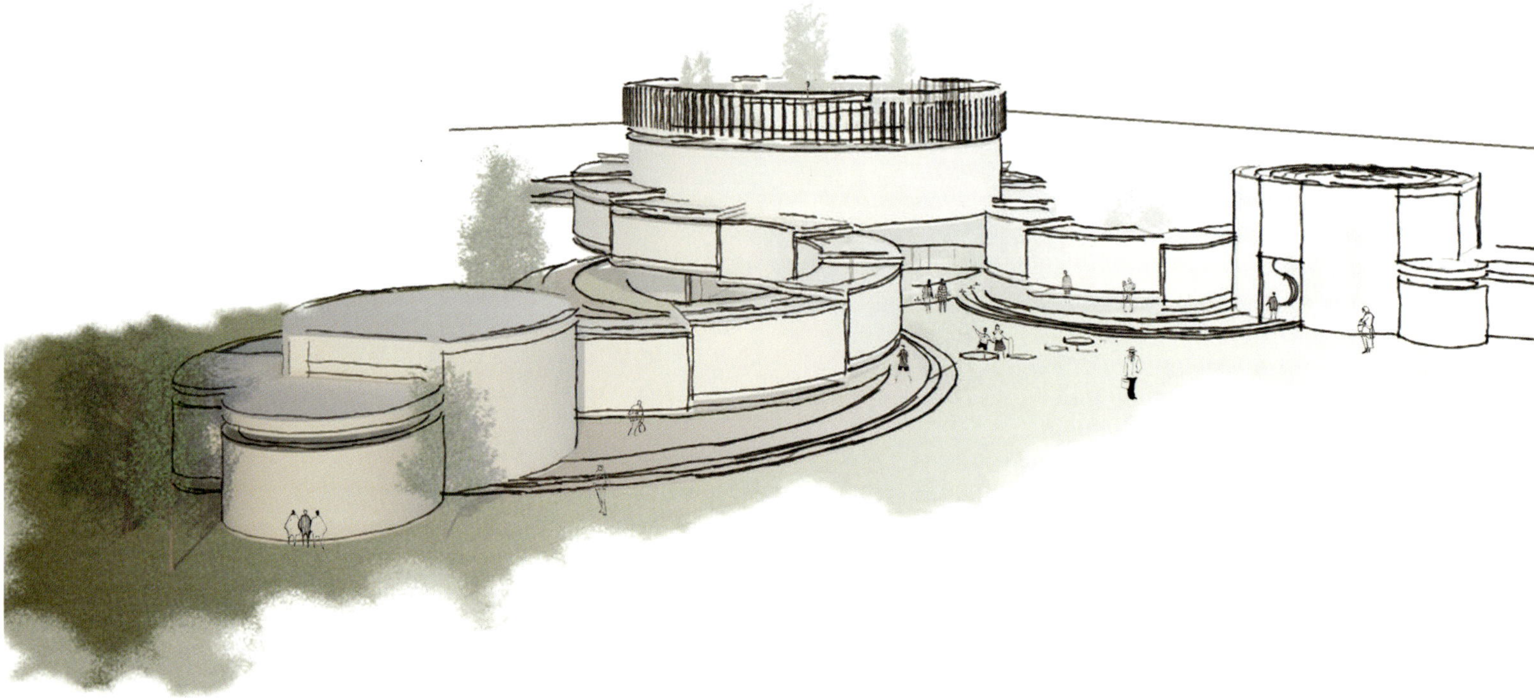

BACKGROUND

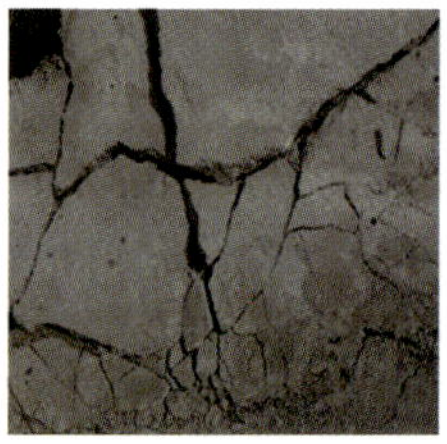

최근 우리나라에도 지진 발생이 많아지면서 지진에 대한 관심이 많아지고 있지만 그럼에도 불구하고 대처 방안이 열악하다. 국내 지진 위험도는 매년 증가하고 실제 피해사례도 나오고 있지만, 건축물을 전혀 제 역할을 못하고 무방비하기 때문에 지진을 중점으로 관련 건물을 설계하였다. 따라서 우리는 지진과 관련된 건물을 설계하는데 중점을 두었다.

Even though Koreans are becoming more interested in earthquakes due to the rapid growth, coping ways are too weak. Although risk the earthquake in Korea is increasing every year and even actual damage cases are reported, buildings are designed to focus on earthquakes because they do not function properly or are vulnerable to earth-quakes. Thus, we focused on designing a building that is related to the earthquakes.

SITE

SITE:난지 한강 공원 잔디 광장
대지 면적: 40117평

한강과 서울은 인구밀집도가 매우 큰 장소임에도 재난에 대한 대피시설이 부족하다. 실제로 서울시의 경우 지진 옥외대피소 중 표지판이 부착된 곳이 거의 없어 시민들이 찾기가 어렵다. 또한 사람들의 대피소 인식 수준도 낮은 편이다. 그래서 우리는 구조물 파손과 낙하물이 없는 한강 공원의 넓은 대지를 활용하고 사람들이 쉽게 대피소를 인지, 파악 할 수 있게 대피소와 전시 · 체험 공간을 함께 구성했다.

SITE: Nanjing Hangang Park Lawn Plaza
LAND AREA: 40117 Average

Han River and Seoul have not enough evacuate facilities, even though those areas are densely populated areas. In fact, some shelters in Seoul do not have signs attached, so it is difficult for citizens to find them. Moreover, people's awareness towards shelters is low. Therefore, we built shelters, exhibition halls and experience spaces using the huge land near the Han River Park, where there are free from damages and falling objects.

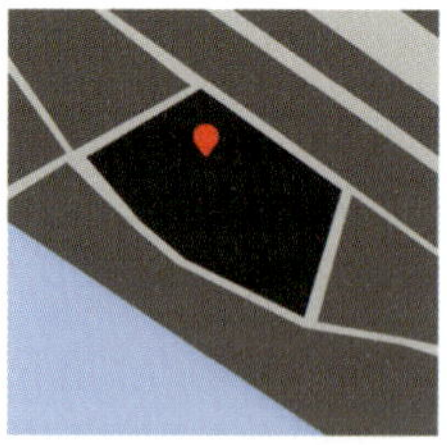

사이트 / Site

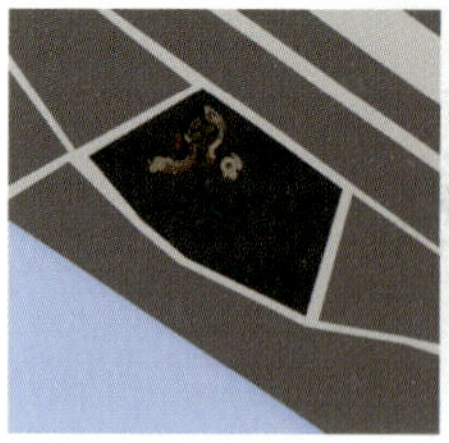

건물 배치도 / layout

주변환경 / Circumstance

접근성 / Accessibility

CONCEPT

징검다리

우리는 '딛다' 라는 단어의 뜻에서 컨셉을 유추하였다. '딛다' 는 두 가지 뜻을 가지고 있는데 첫 번째는 '어려운 상황 따위를 이겨내다' 두 번째는 '발을 올려놓고 서거나 내리누르다' 이다.
첫 번째 뜻은 재난을 무거운 상황으로만 생각하지 않고 엔터테이먼트적인 요소들로 극복해 나감을 의미하고 체험관으로 치환되며 두 번째 뜻은 재난 상황 속에서 대피소에 발을 딛고 있음으로써 그 상황을 극복해 나감을 의미하며 대피공간으로 치환된다. 그래서 우리는 평소에는 재난 간접 체험 및 대피소 체험장으로 사용하다가 재난 시 대피소로 사용하는 가변형적 건물을 설계했고 '징검다리' 라는 컨셉을 생각해냈다.

We inferred our concept from the word ' to step on'. It has two meanings. The first one is to 'overcome difficult situations' and the second one is to 'stand with your feet up or down'. The first meaning implicates that disaster is being overcome by entertainment factors and is replaced by experience rather than considering it too deep. And the second implication is by the stand at the evacuation site in a disaster situation and overcome it and substituted with a place of evacuation. Therefore, we devised the concept of 'stepping stone'. and designed a building, which can be used as a disaster shelter from indirect experience and experience shelter.

PROCESS

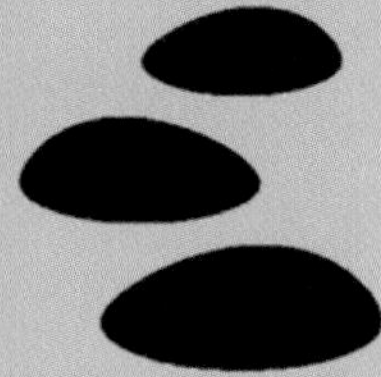

1. 형태

징검다리의 돌과 물의 형태로 돌은 건물, 물을 통로로 흐르는 듯한 디자인을 하였다.

1. Formation

We defined a form of stone and water, which are designed like flowing water or passages.

2. 파장

섬돌에 발을 딛으면 물과 돌이 밟혀 생기는 파동을 내부 조닝, 발코니, 천장에 디자인하였다.

2. Wavelength

When you step on the stone and the water, the wavelength that has been created, will be designed directly to the internal zoning, balconies, and ceiling.

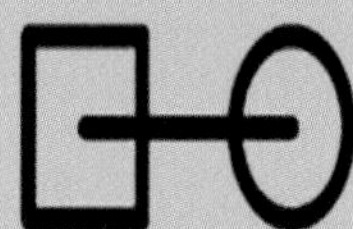

3. 연결

징검다리로 두 개의 땅이 연결되는 모습을 따와서 통로로 연결되는 세 건물. B건물 1층 평면에 내부와 외부의 관입으로 디자인하였다.

3. Connection

We then designed the building to be entered into three connected buildings - to the first floor of the building B and two sections that lead to the hallway.

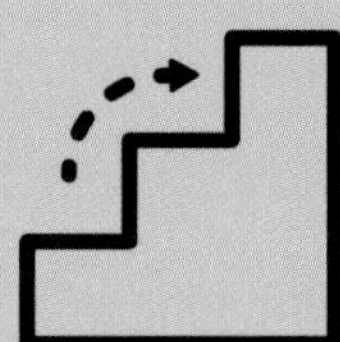

4. 단계

변화에 따른 단계와 동선에 따른 단계로 크게 두 가지로 나뉜다. 변화로는 높낮이에 단계를 줘서 수직적으로는 건물과 통로의 높이 차이를 형성하고, 수평적으로는 건물과 발코니, 데크의 면적을 다르게 하였다. 동선으로는 A–B–C 건물 순으로 체험을 진행하여 순서를 만들어 단계를 형성하였다.

4. Stage

The stage will be divided into two types according to transformation and to a moving line. The height changes indicate the vertical difference between the building and its passages. The area of buildings, balconies, and decks varies horizontally. On the moving line, the A-B-C building is built in the order of experience, and a sequence for forming the stairs is created.

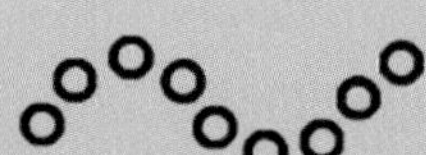

5. 경계

징검다리로 인해 물살에 생기는 경계선에서 모습을 따와 통로의 갤러리 부분과 복도 부분의 창과 빛의 양의 차이로 경계를 디자인하였다.

5. Boundary

We designed the boundary by the difference between the number of windows and the portion of the passage that appears at the water-formed boundary due to the light in the gallery portion of the aisle and the stair bridge.

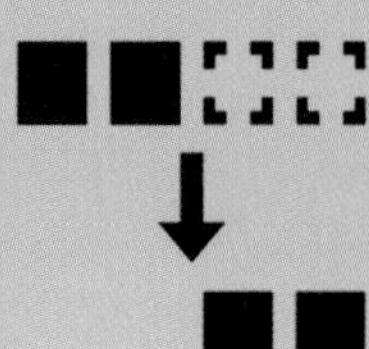

6. 분리와 연결

징검다리는 끊어짐과 동시에 연결되어 있다는 역설적인 모습을 가지고 있다. 통로로 이어지면서도 끊어진 느낌을 내기위해 부분적으로 통유리로 디자인 하였다. 창이 적은 통로에서 색다른 느낌을 낼 수 있는 공간이다.

6. Separation and connection

The stepping stone has an ironic twist in the staircase and it is connected together with the break. We designed it partially to feel disconnected as we walked down the aisle. Then We designed it partially to feel the disconnection as we walked down the aisle. Here is the place where you can feel a different feeling with the few windows.

7. 시퀀스

건물들과 갤러리의 기능을 구분해 인지, 대피, 대처 순서에 맞춰 공간을 구성하였다.

7. Sequence

We designed the space depends on its functions, depending on the recognition, evacuation and coping order of the building and gallery.

MASS STUDY

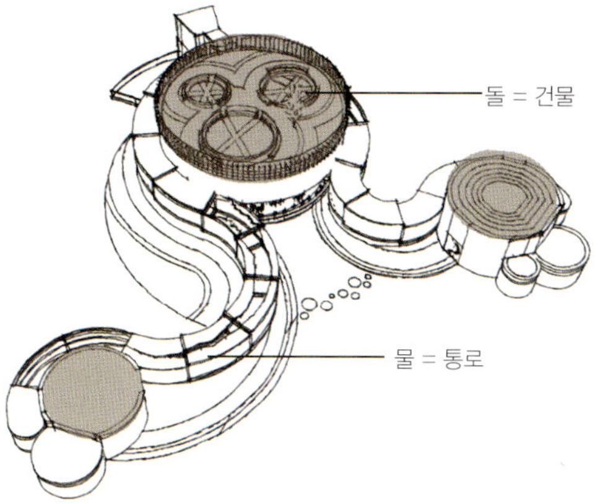

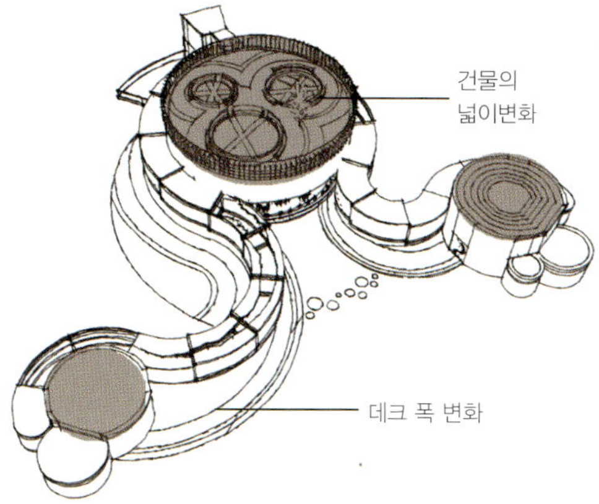

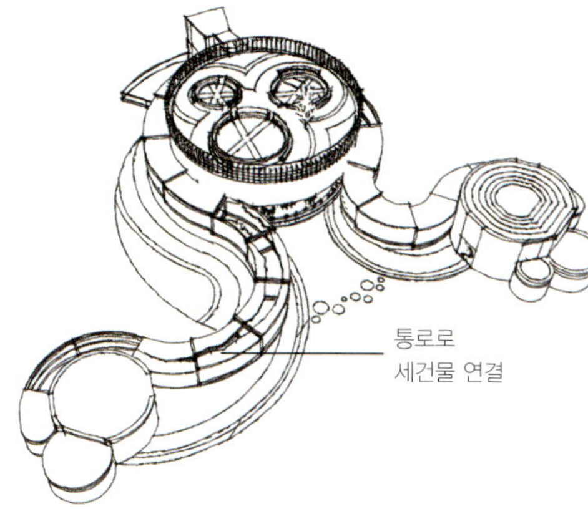

ISOMETRIC

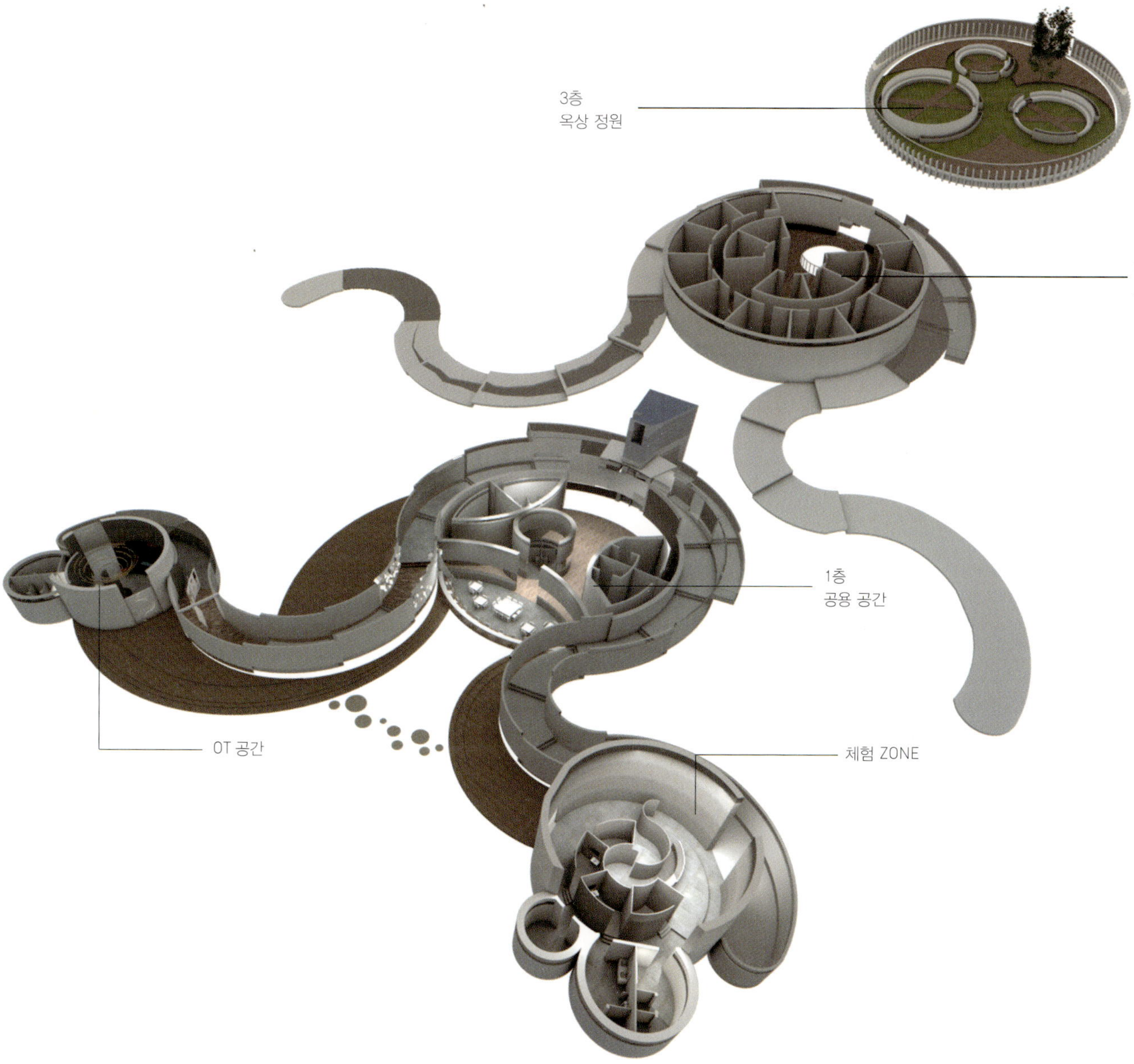

PERESPECTIVE

- A 건물 : OT 공간

체험 시작 전 재난 (지진)에 대해 공부하고 체험에 대해 설명듣는 곳.

- A building : OT Space

The area where we preview about earthquakes before the actual experience.

- C건물 : 체험 ZONE

지금까지 보고 들은 내용을 현실감있게 체험해 보는 곳. 지진 발생 전 부터 지진 발생 후 대피소 체험까지 가상현실로 체험해 보는 곳.

- C building : Experience Zone

The area where we experience realistically from what we've heard from now on. The place to experience virtual reality from before an earthquake to after an earthquake to an evacuation center.

- B건물 : 1층 식당

B건물은 인간이 사는데 꼭 필요한 것이 자는 것, 먹는 것, 씻는 것이라고 생각해서 주방, 화장실, 샤워실로 이루어진 공용공간.

- B building : 1F Restaurant

Building B is a public space consisting of kitchen, bathroom, and what it is necessary for people to live in.

- B건물 : 3층 옥상

옥상 정원.

- B building : 3F ROOF TOP

ROOF GARDEN.

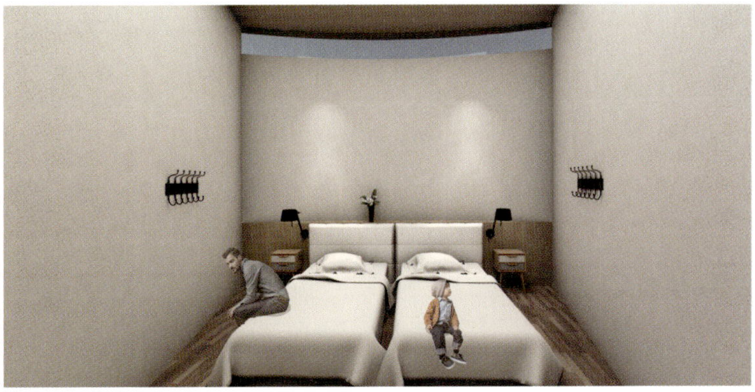

- B건물 : 2층 거실

작은 거실, 공용 공간, 1층과 2층의 연결.

- B building : 2F living room

The small living room, a public place-Connection of the first floor and the second.

- B건물 : 2층 방

취침 공간, 개인 공간.

- B building : 2F room

Sleeping space. Personal space

PERESPECTIVE

– 갤러리

A 갤러리
인지의 공간으로 지진에 대해 알려주고 강도별 지진 피해 상황을 시각적으로 4단계로 표현하여 갤러리를 지나가면서 지진에 대해 인지한다.

- Gallery

A Gallery
We will aware of the earthquake as a space of awareness, expressing the damage of the earthquake in four visual stages, and recognizing earthquakes as we go through the gallery.

– 갤러리

B 갤러리
대피의 공간으로 대피소를 구경하고 대피소 내진설계에 대해 전시되어있다.

- Gallery

B Gallery
We look around the evacuation center and it is displayed an earthquake-resistant design.

– 갤러리

C 갤러리
대처의 공간으로 지진 후 가장 큰 피해인 트라우마에 대한 공간이다. 트라우마의 가장 큰 대처는 공감과 예방으로 통로부분에는 과거에 지진 트라우마를 겪고 이겨낸 사람들의 이야기들을 전시했고 예방하기위해 체험zone으로 넘어간다.

- Gallery

C Gallery
It is a space for the biggest traumas after the earthquake. We placed stories of people who have overcome the effects of an earthquake in the past at the hallway and we are to go through the experience zone to prevent earthquakes.

– 아이 레벨

외부.

- Eye Level

Outside.

PLAN

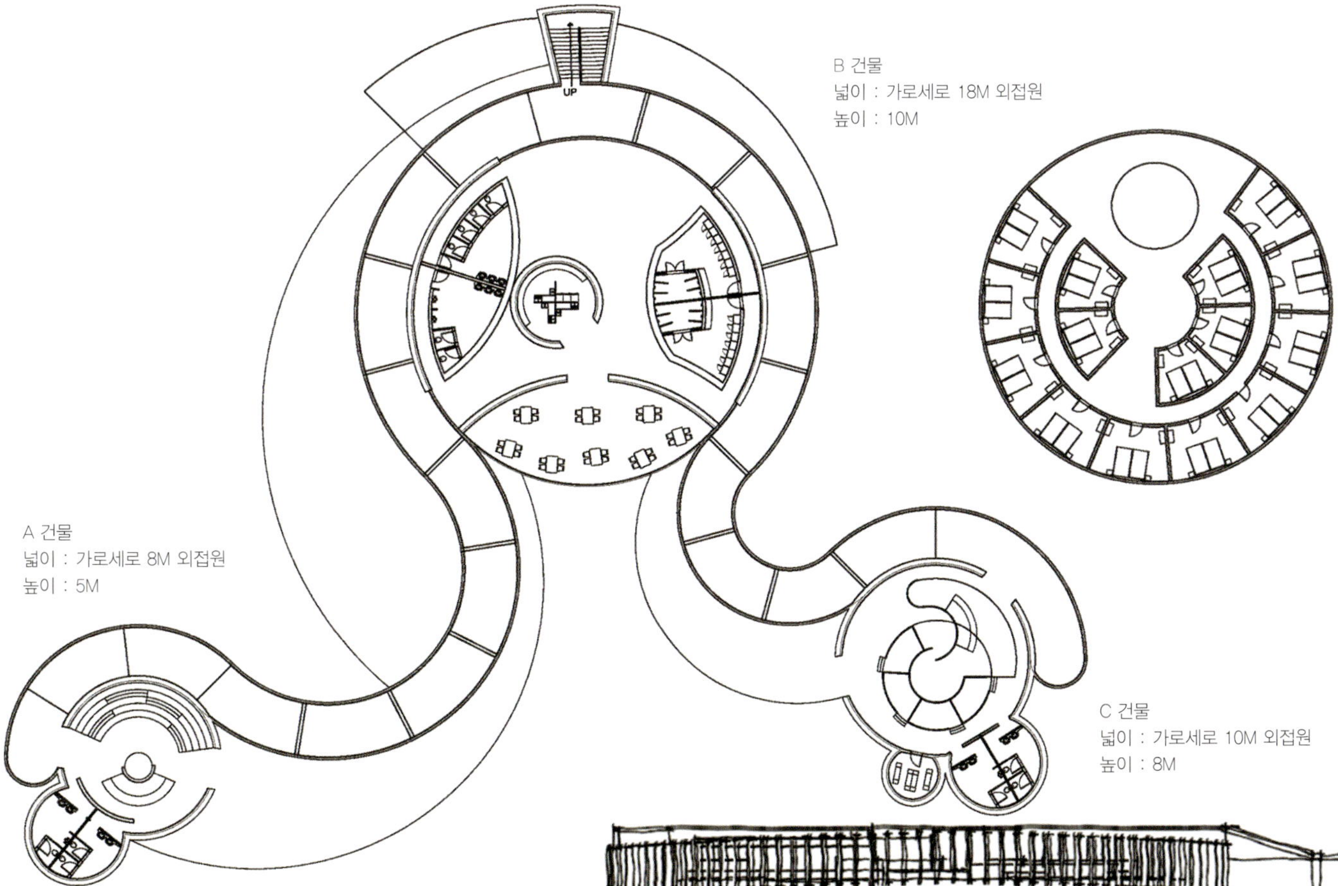

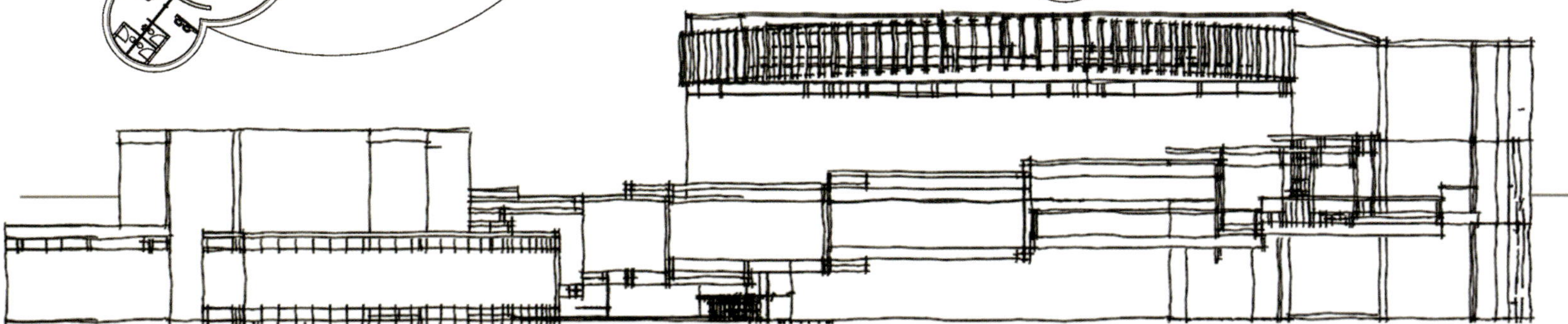

조정민 I Cho jeong min 고은혜 I Ko eun hye 문소연 I Moon so yeon

REVIVE

기존의 것을 재생시켜 새로운 공간으로 되살리다

Revitalize existing space and bring it back to a new space

BACKGROUND

전쟁을 피해 피난 온 남수단의 이재민과 아이들로 가득한 우간다의 난민촌 '비디비디'. 내전을 피해 우간다로 탈출한 난민수가 100만 명을 넘어섰고, 역사상 최대 규모의 단일 난민 유입 상태이다. 매일 100여명이 넘는 아이들이 총격을 피해 홀로 국경을 넘고 있으며, 그렇게 덩그러니 세상의 경계로 내몰리는 상황이다. 식량, 식수, 의약품 등 모든 것이 턱없이 부족한 열악한 환경 속에서 웃음을 잃어버린 아이들... 누구의 보호도 받지 못하고 폭력을 겪거나 각종 범죄에 노출되는 아이들에게 가장 필요한 것은 '일상' 이다. 그래서 아이들이 공부하고, 웃고, 뛰 놀며 평범한 일상으로 복귀할 수 있도록 그에 맞는 환경과 기회를 제공해야 한다. 삭막한 난민촌에 유일하게 '정상적인' 건물과 놀이터에서 아이들이 공부하며 웃음이 쌓이는 공간에서 그들의 깊은 상처를 치유해 준다.

There is a Uganda's refugee camp called 'BidiBidi'. It is filled with refugees and children from South Sudan, who escaped from the war. Bidi Bidi has been the single largest refugee in history after fleeing to Uganda. More than 100 children cross the border alone every day to avoid shooting, and that's how they were pushed to the edge of the world. Children who have lost laughter in poor conditions, such as food, water, and medicines are too scarce, are most in need of children who go through violence any protection. Therefore, it is necessary to provide them the appropriate environment and opportunities, such as study, laugh, play, and be branded as normal. Only in the stark refugee camp could make children study in " normal " buildings and playgrounds to heal their deep wounds in laughter.

SITE

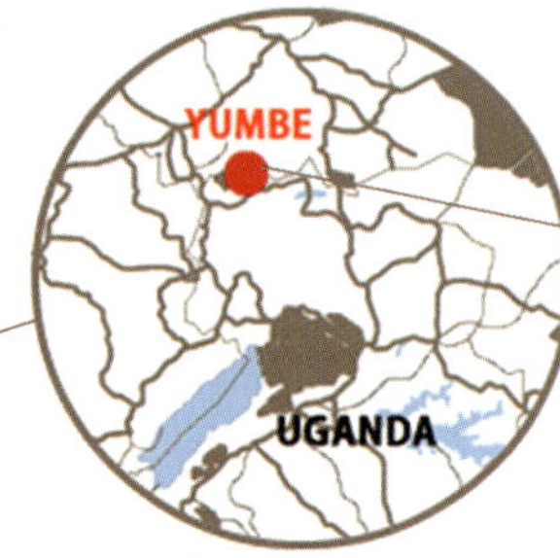

YUMBE

BIDIBIDI

SITE ANALYSIS

우간다의 남수단 난민촌

우간다 '비디 비디' 난민촌

세계에서 가장 어린 나라, 내전의 고통에 신음하는 나라, 그러나 누구도 주목하지 않는 나라, 남수단. 2013년부터 지금까지 이 죽음의 땅을 피해 남수단 난민 약 90만 명이 우간다로 도피했다. 그 중에서도 우간다 비디비디(Bidibidi)는 전 세계에서 가장 많은 난민이 살고 있는 세계 최대 난민 정착촌이다. 비디비디에 정착한 난민 중 무려 68%가 18세 미만 아동인 현실이다.

South Sudan is the youngest Country in the world. So many citizens are suffering from a civil war, but nobody notices it. Since 2013, nearly 900,000 South Sudan refugees have fled to Uganda to escape their deadly land. Uganda's Bidibidi is the world's largest refugee settlements with the largest number of refugees in the world. As many as 68 % of refugees have settled and most of them are children under the age of 18.

272,206명 Bidibidi > 161,661명 Palloliya > 107,330명 Imbepi > 86,770명 Reno

→ 우간다 내 정착촌에 있는 난민수

683,160명

2016년 이후 우간다로 온
남수단 난민수

950,562명

2017년 현재 우간다에 정착한
남수단 난민수

86%

전체 난민중
여성, 아동 비율

TARGET

우간다 난민촌 인근의 아동난민과 소외된 취약계층의 아이들

계속되는 분쟁과 폭력으로 인해 교육의 권리를 박탈당한 아이들에게 책에서의 배움(learn)을 통한 학습의 기회를 제공하고, 그들이 죽을 때까지 경험해 보지 못하는 신비로운 환경을 공간에 실현시켜 놀이(play)와 체험(experience)을 통해 잠시나마 전쟁의 아픔을 이겨 낼 수 있는 즐거운 치료와 소통의 공간이 되었으면 한다. 더불어 이런 공간이 제공됨으로써 공통된 사연을 가진 그들이 서로 의지하며 교감이 나누는 환경이 조성되기를 바란다.

For the children who were educationally deprived from the continuous trouble and violation, I hope to provide the "Learning" opportunity from the book and by realizing the mysterious environment to the space, through the play and experience, i sincerely hope the place where those things can lead to them to endure the pain of war for a while. And also, by providing 'this' environment, i hope that people who have general stories depend each other and share their feelings

CONCEPT

개선, 재생시키다.

실제 버려진 우간다의 교회건물을 크게 훼손하지 않는 범위 내에서 recycling하고 재료 또한 폐자재를 upcycling하여 어린이 들을 위한 공간으로 재해석하는 일종의 재생프로젝트인 셈이다.

Improve, regenerate.

It is a kind of recycling project that recycles materials that do not seriously undermine the abandoned Ugandan churches and recycles materials as well as upcycling waste materials into space for children.

도서관 & 열람실	놀이 & 체험	외부 놀이시설

PLAY LIBRARY

놀이공간과 열람실이 결합된 도서관
A library that combines a playroom and a reading room

앞쪽은 기존 도서관의 사전적 형태와는 다르게 자유로운 형태와 활동적인 분위기로 아이들이 거부감 없이 친근하게 드나들 수 있게 디자인했다. 안쪽의 체험과 놀이공간은 그들에게는 신비롭고 무지한, 어떻게 보면 두려울 수도 있는 우주라는 공간을 녹여 인상적인 체험과 함께 잊지 못 할 흥미로운 기억을 심어준다. 한마디로 각기 다른 주제를 강조하며 공간 영역에 호기심을 불러 일으키는 것이다. 이렇게 내부의 두 공간을 열람실과 놀이시설의 형태로 명확하게 구획해 분리시켜 놓았지만, 외부를 통해 두 공간이 자연스러운 흐름으로 연결되게 했다.

In the front, unlike the dictionary form of the existing library, it was designed in a free form and an active atmosphere so that the children could entertain without intrusion. The inner experience and play space gives them a memorable space of space that can be mysterious, ignorant, scary in some ways, giving them memorable experiences and an unforgettable memorable memory. In a nutshell, it emphasizes different subjects and causes curiosity in the spatial domain. In this way, the two spaces in the interior are clearly separated and separated in the form of a reading room and a play facility, but the two spaces are connected by a natural flow through the outside.

MATERIAL

UPCYCLING / RECYCLING (cement pipe, tire, pallet, steel)

전체적으로 적은 비용으로 빠르게 공급될 수 있는 폐 재료들로 UPCYCLEING하였고 현지에 있는 재료를 최대한 활용하였다. 또한 각 공간의 컨셉에 맞춰 재료를 달리 함으로써 교통의 흐름과 실내외 공간의 변형이 전개 될 수 있도록 했고 폐 재료들을 가공하지 않고 본연의 상태를 그대로 가져와 러프하고 무질서한 분위기를 끌어냈다.

Generally, We used waste materials, which could be supplied quickly and at a lower cost. Moreover, local materials have been utilized to the maximum. In addition, materials were distributed differently based on the concept of each space and place, allowing for the flow of traffic and changes in indoor and outdoor spaces. Instead of processing waste materials, we decided to bring it to the original state to life, trying to creating some mess naturally.

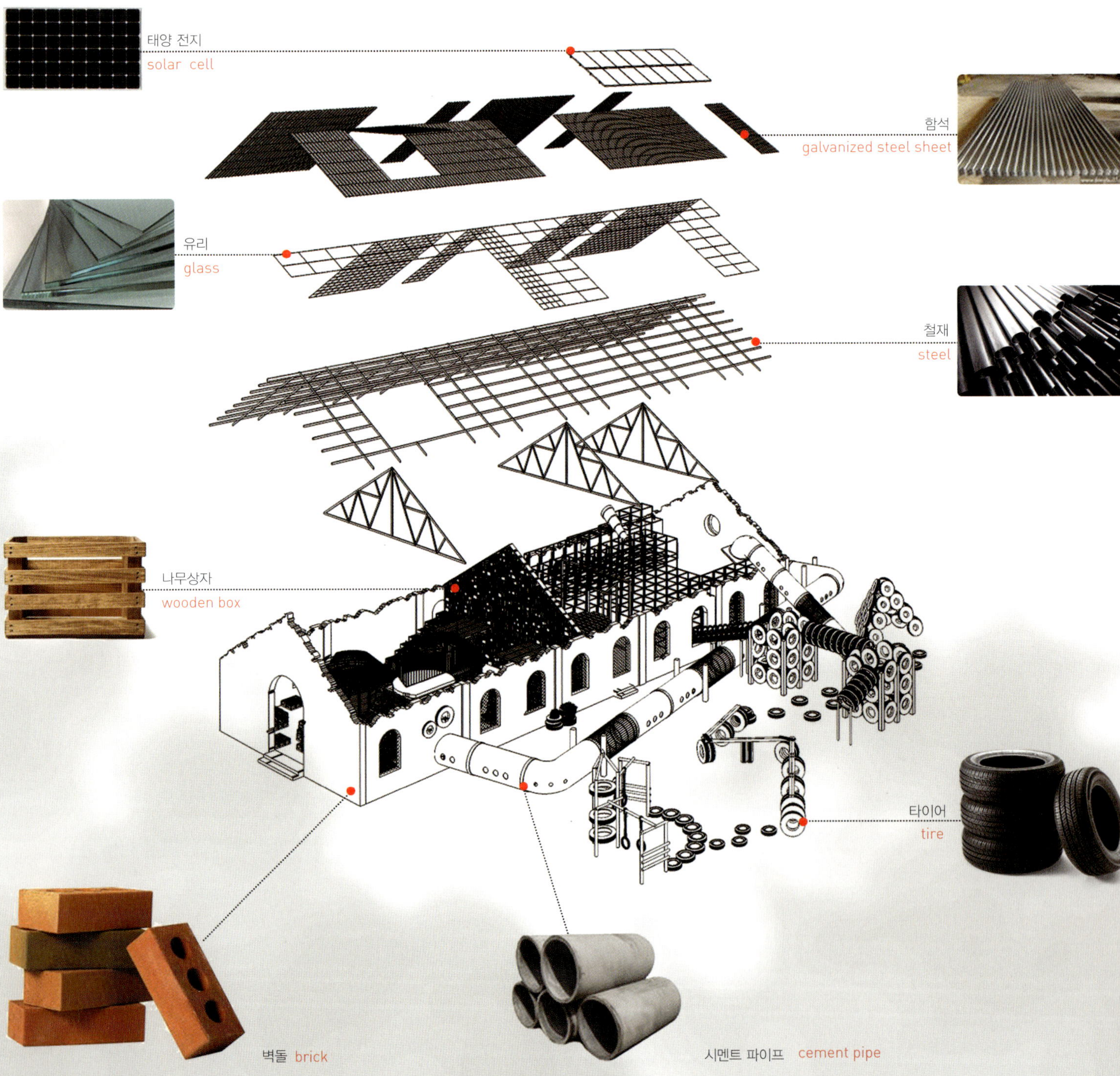

ISOMETRIC

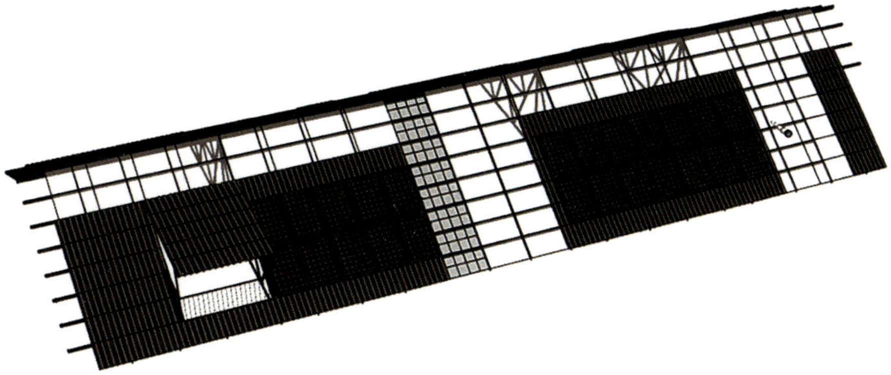

● Section-Roof

버려져있던 건물의 지붕구조를 그대로 가져와 재활용하다.
자연의 에너지로 전기를 얻을 수 있는 태양열과, 더운 곳이니만큼 바람이 잘들 수 있는 테라스, 햇빛 가림막 캐노피를 만들었다.

Taking and recycling the roof structure from abandoned buildings. As it is hot and sunny with solar power, it has a terrace, sunbeam, and canopy to let people get electricity from nature.

● Section-Reading Room 열람실

나무상자로 쌓아올린 책장은 가벽의 기능과 동시에 놀이공간으로 넘어가는 문의 역할을 한다. 책장 앞에 계단과 같이 pallet를 층층이 쌓아 높은 곳에 있는 책을 꺼내 볼 수 있게 했다.

This wooden bookcase that is stacked with wooden boxes serves as a door to play along with the wall. We put a small ladder in front of the bookcase for you to take books from the top.

아이들이 책을 읽는 공간은 pallet를 재활용해 움집 형태로 재해석 했고, 바닥에 pallet를 배치해 아이들이 앉아서 책을 읽거나 징검다리처럼 건너다닐 수 있게 했다.

Childrens' reading room was recycled and reinterpreted into a mud hut. And the board was placed on the floor so that the children could sit down and read. Those boards are also showing the design of stepping stones.

● Section-Terrace 테라스

바깥을 보면서 바람과 햇볕을 느끼며 쉴 수 있는 공간

Place where people can take some rest with the nice view

● Section-Play ground 놀이공간

아동난민촌에 이질적일 수 있는 우주공간을 정글짐이라는 놀이시설로 표현하였고, 내외부가 연결된 정글짐을 아이들이 오르내리며 놀 수 있는 공간으로 꾸며 보았다.

This place was described as an amusement facility called " jungle gym, " and the inner and outer of the jungle gym was decorated with spaces for children to play with.

또한 정글짐을 타고 올라가면 외부로 연결된 미끄럼틀을 통해 열람실로 이동이 가능하다.

Also, when you ride up a jungle gym, you can directly go to the reading room through a slide, which is c

PROGRAM

Link

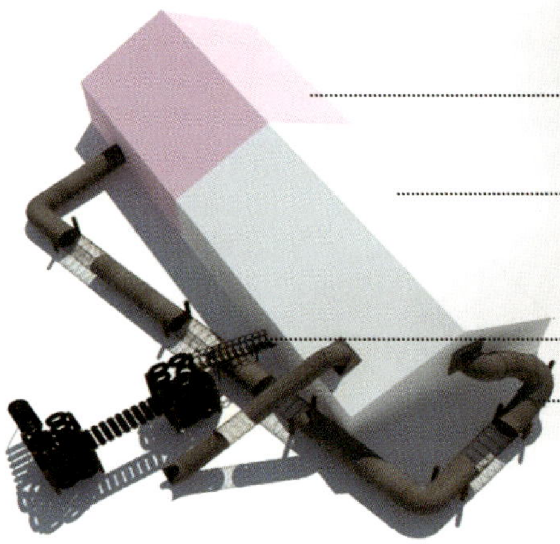

두 공간을 맺어주는 매개체 역할을 한다

It serves as a bridge between two places.

매개체1 : 열람실과 내부 놀이시설을 연결

Instrument 1 connects reading room and playing facilities inside

매개체2 : 내부 놀이시설과 외부 놀이시설을 연결

Instrument 2 connects inside and outside play facilities

Multifunction

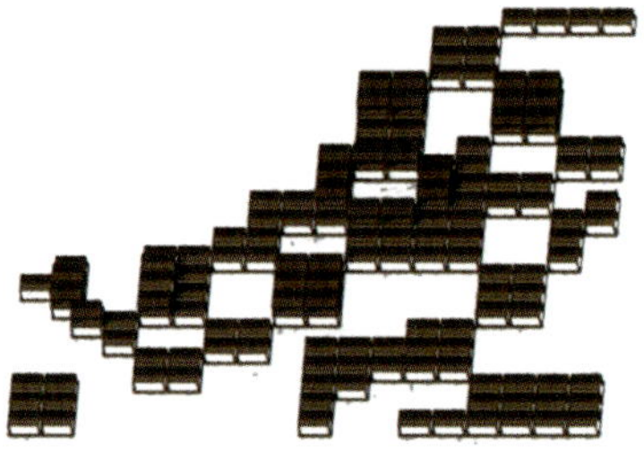

바닥에 팔레트를 불규칙적으로 깔아 리듬감이 있는 징검다리 형태의 이동수단으로 재탄생 시켰고, 더불어 아이들이 앉아서 책을 볼 수 있는 의자의 기능을 부여하였다.

Palettes were randomly laid on the floor and let it seen as a rhythmic stepping-stone vehicle, with the function of a chair for children to sit on and read.

팔레트를 규칙적으로 쌓아 단을 주어 계단의 기능을 하면서 높은 책장의 책을 직접 꺼낼 수 있는 사다리의 역할도 한다. 또한 여럿이 앉아 쉴 수 있는 휴식의 공간도 될 수 있다.

Those are also piled up regularly to serve as a ladder to pull out books on the high bookcase directly, serving as a staircase. It can also be a relaxing space for people to sit down.

벽을 가득 채운 나무상자는 책을 진열하는 책장의 기능을 가지며 열람실과 내부 놀이 시설을 분리시켜주는 가벽의 역할을 가진다.벽을 가득 채운 나무상자는 책을 진열하는 책장의 기능을 가지며 열람실과 내부 놀이 시설을 분리시켜주는 가벽의 역할을 가진다.

Wooden boxes have a role as a bookcase for displaying books and serve as a temporary wall separating the reading room from the inner play area.

다양한 형태로 변형되어 다목적 기능을 가진다

It is modified in various forms and has a multipurpose function.

PLAN

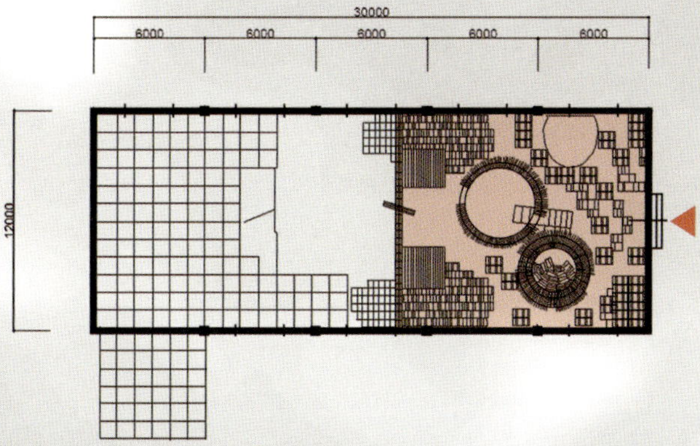

열람실

도서관으로서의 기능은 기본적으로 갖고 있으나, 기존의 정적이고 딱딱한 도서관의 구조와는 확연히 다른 컨셉을 잡았다. 자유로운 구도에 동적인 상황, 움집으로 스토리를 정하였고, 징검다리 처럼 바닥에 파레트로 단을 올렸다.

Reading Room

Although it has basic functions as a library, it has a distinct concept from the traditional structure of the library. We set a dynamic house with a free structure and placed a few palettes on the floor as if they were stepping stone.

PLAN

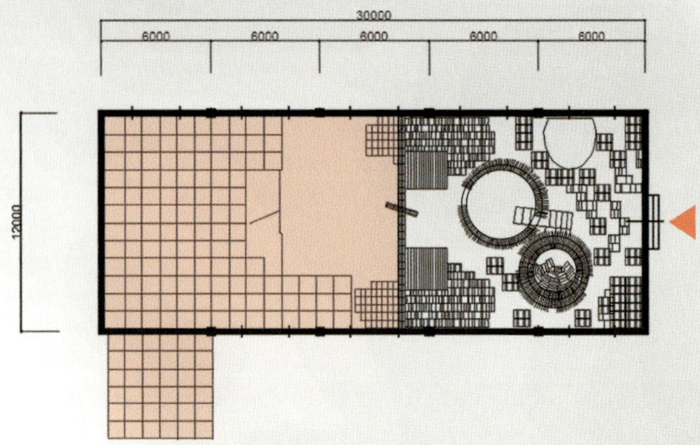

놀이시설

가벽을 지나 놀이시설로 들어오면 앞쪽의 도서관과는 완전히 다른 컨셉의 공간이 나타난다. 아동난민촌에 이질적일 수 있는 우주를 아이들에게 재미있게 풀어 낼 수 있는 공간 컨셉으로 디자인하였으며, 사람의 육체에 중력이 작용하지 않아 힘이 통하지 않는다는 우주라는 개념은 힘의 억압아래 있는 아이들에게 해방감의 상징을 부여한다.

Playing Facilities

As we pass the wall and enter the playing facility, you will see a completely different concept of place from the library in front. We designed it under the concept of a space.

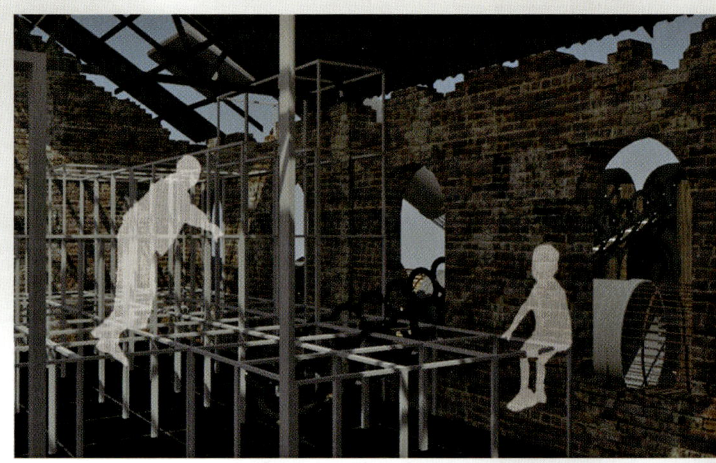

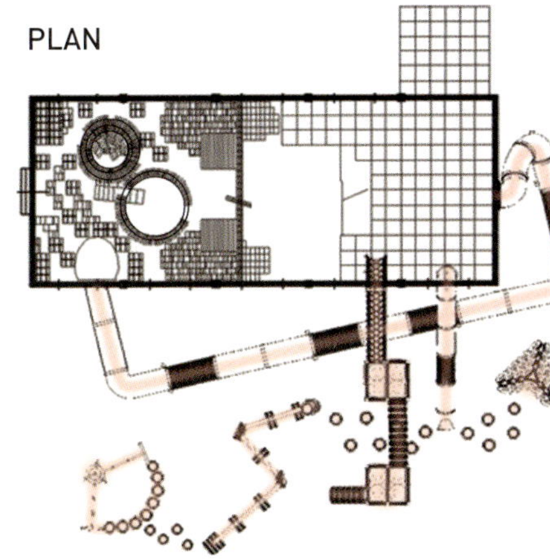

외부 놀이시설

놀이공간의 우주컨셉을 외부로 자연스럽게 이끌어 두 공간은 분리되어 있지만 하나의 매개체로 상호관계를 이루게 디자인했다. 하지만 외부공간을 이루는 재료를 타이어와 시멘트 파이프 등으로 차이를 주어 내부와는 다른 광범위하고 유기적인 형태의 우주를 표현하였다. 시멘트 파이프를 긴 통로 형태로 연결시켜 놀이시설부터 열람실까지의 두 공간이 내부에서는 철저히 나누어져 있지만 외부를 통해 간접적으로 연결되게 했다.

Outside playing facilities

The concept of the universe in the theme park is naturally separated, but they are designed to be correlated with one place.

Atmospheric Cloud

: 대기 중 오염된 구름

재난의 종류들 중 우리는 현재 눈에 보이지 않지만 축적되어 발생하는 인위적 재난인 대기오염을 선택하였다. 땅은 작지만 인구밀도가 높은 우리나라는 다른 국가들에 비해 에너지 사용률이 높다. 그 만큼 이산화탄소의 배출량과 매연, 가스들에 의해 대기오염이 심각하게 일어나고 있는 상황이다. 이런 재난은 사람들이 사태의 심각성을 충분히 인지하지 못한다. 그렇기에 우리는 이 심각성을 알릴 수 있도록 하나의 상징적인 건물을 통하여 전시 체험 공간을 선보일 것이다.

The theme of this International exchange exhibition is a 'disasters'. We chose the atmospheric pollution among the different types of disasters, which is an artificial disaster that is invisible but accumulative. Though South Korea is smaller but densely populated than other countries, has a high energy use rate. Therefore, Carbon dioxide emissions, soot, and gases are causing serious air pollution. What is important is that people do not realize the seriousness of the air pollution. Thus, we would show this seriousness by showing the exhibition place through a symbolic building.

조희연 | Jo Heeyeon, 이정민 | Lee Jungmin

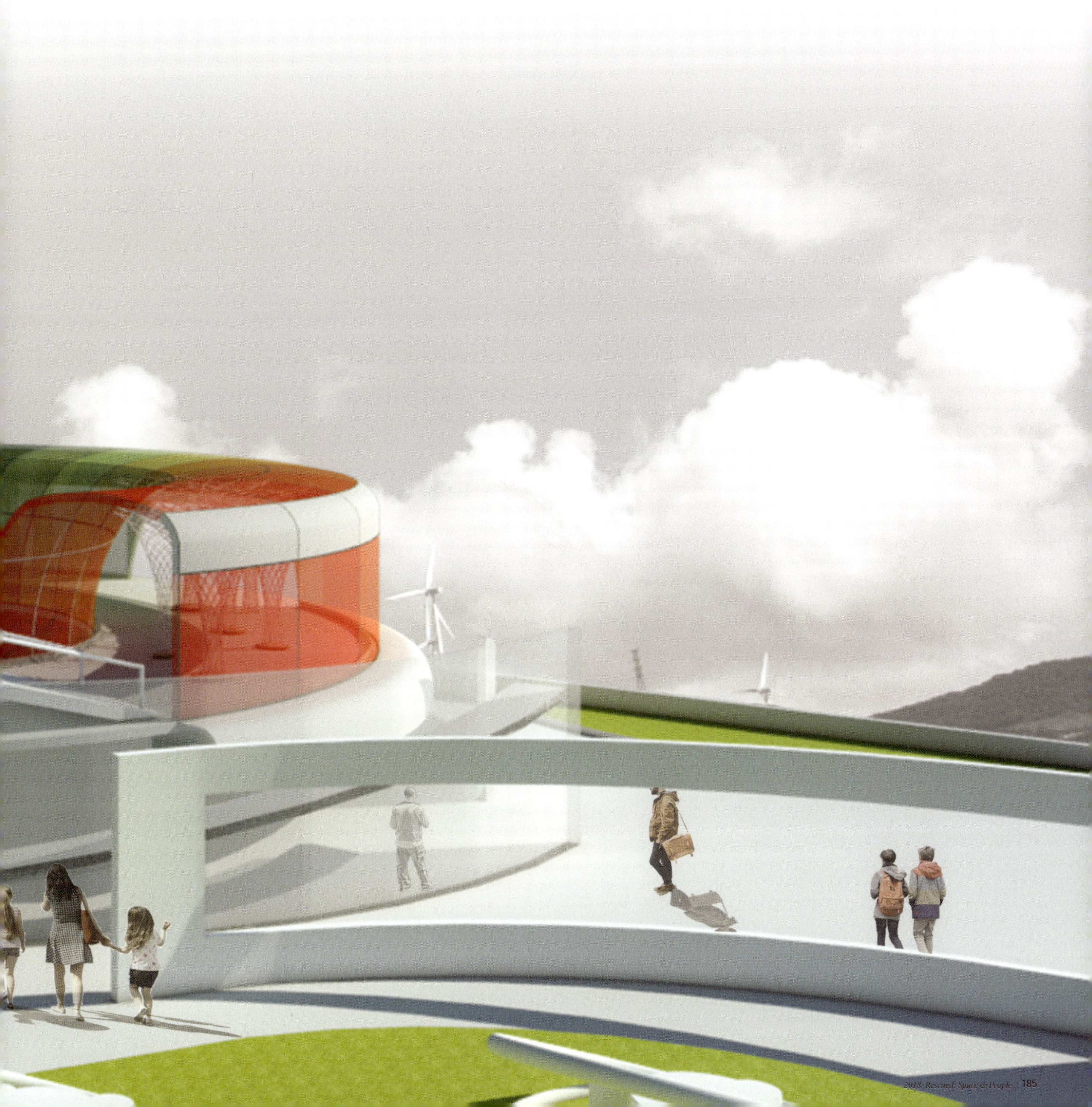

SITE

서울특별시 용산구 한남동에 있는 뮤지컬 대중음악 전용 공연장 블루스퀘어.

서울 용산구 이태원로 294 블루스퀘어
3,150㎡ = 약 952평

Blue Square, musical performance venue dedicated to music in Hannam - dong, Yongsan - gu, Seoul.

294 Blue Square, Itaewon Road, Yongsan-gu, Seoul
3,150㎡ = about 952 pyung

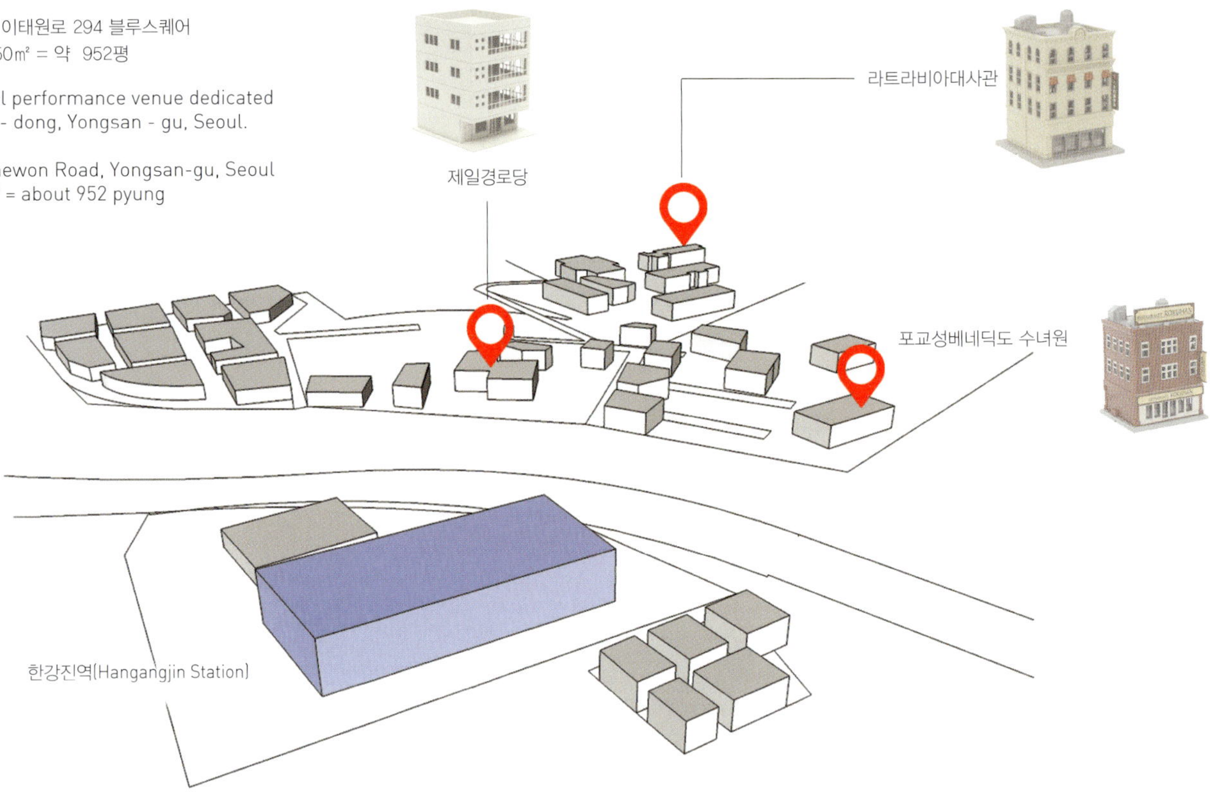

BACKGROUND

서울은 영국 파이낸셜 타임즈에서 선정한 전세계에서 대기오염이 가장 심한 도시 TOP 3에 선정되었다. 그러나 현재 서울에 살고 있는 사람들은 바쁜 생활 패턴으로 대기오염에 대한 심각성을 인지하지 못하고 있다. 높은 건물들로 인해 하늘은 가려져 있고, 스마트폰의 사용으로 고개를 숙인 채, 땅만 바라보며 살아간다. 현재 하늘은 원래의 색을 잃은 채 오염되어 있다.

대기오염 중 제일 문제가 되는 미세먼지는 1급 발암물질로 몸 속에 축적될 경우 우리의 생명의 위험과도 직결된다. 국제 학술지 네이처에서는 초미세먼지의 영향으로 한국과 일본의 조기 사망자가 연간 3만명에 달한다고 발표했으며, OECD(경제협력개발기구)는 대한민국의 대기오염이 개선되지 않는다면 2060년까지 한국인 900만명이 조기 사망할 수 있다고 말한다.

Seoul was selected by the British Financial Times as the top three cities in the world with the highest levels of air pollution. However, people currently living in Seoul are not aware of the seriousness of air pollution in their busy life patterns. Due to the high buildings, the sky is hidden, and I look down on the ground while using my smartphone. The sky is now polluted with the loss of its original color.

The first problem of air pollution, fine dust, is accumulated as a first-level carcinogen in our body, which directly affects our life risk. According to the International Journals Nature, which is highly credible globally, OECD (Organization for Economic Co-operation.

SITE ANALYSIS

전시공간은 유동인구가 많은 서울, 그 중 한남동에 위치한 뮤지컬 대중음악 공연장인 블루스퀘어이다. 다양한 연령대가 오고 가는 문화공간으로써 옥상에 서울 시내를 내려다 볼 수 있는 전망을 가지고 있다. 그렇기에 외부에서도 한눈에 전시공간을 볼 수 있어 대기오염이라는 주제를 사람들에게 인식시키기에 용의하다.

The exhibition space is Seoul, which has a large population of people, and Blue Square, which is a musical popular music venue located in Hannam-dong. It is a cultural space for various ages, and has a rooftop view overlooking Seoul city. Therefore, we can see the exhibition space at a glance from the outside, and are willing to make people aware of the theme of air pollution.

PERESPECTIVE

환경문제에 대한 인식. 오랫동안 축적되어 발생하기 때문에 사태의 심각성을 사람들이 인지하지 못하고, 사건이 터지면 인지하나, 다시 원점으로 돌아간다는 점이 있다.

지구온난화 -> 인위적 재난

이러한 환경문제를 사람들에게 인식시키려면 순간적 경험이 아닌 지속적인 경험이 필요하다는 결과가 나왔고, 단순히 전시를 보고, 체험하는 곳이 아닌 하나의 상징성이 있는 건물로 인식시켜 보러오는 개념을 생각해 내었다.

Recognition of environmental problems.
Because it is accumulated over a long period of time, people can not recognize the seriousness of the situation, but if the event happens, it returns to the origin.

Global Warming --> Artificial Disaster

To recognize these environmental problems, we came up with the idea that we need constant experience, not momentary experience. We came up with the idea of seeing the exhibition and seeing it as a symbolic building rather than a place to experience it.

MOTIVE

산성비, 오존층 파괴, 온실효과, 스모그 현상의 공통적인 부분의 원인은 대기 오염 물질이라는 것이다. 그 중 우리는 우리가 만들어낸 오염물질이 대기 중에 피어오르는 현상을 모티브로 구성하게 되었다.

The common cause of acid rain, destruction of the ozone layer, greenhouse effect, and smog phenomenon is attributed to air pollutants. Among them, we have made up the motif of the phenomenon that the contaminants that we created rise in the atmosphere.

PIAN SPACE

1F
진입로 : 공원
안개의 공간 : 재난의 현상
빛의 공간 : 재난의 인식

2F
바람의 공간 : 재난의 극복
테라스 : 휴식 공간
램프(전망대) : 공기의 순환

1F
Driveway: Park
Space of mist: The phenomenon of disaster
Space of Light: Recognition of Disaster

2F
Space of the Wind: Overcoming Disasters
Terrace: seating area
Lamp (Observatory): Circulation of air

1st FLOOR PLAN

중심 건물을 기준으로 흐름에 따라 가벽을 설치하여 균형을 잡아준다. 이 가벽은 '대기 층' 을 뜻하며 유기적인 곡선으로 미로의 시퀀스 공간이 되기도 한다.

Based on the central building Install a wall to balance. The symbol of this wall is the 'atmospheric layer'Organic curves can also be a sequence space for mazes.

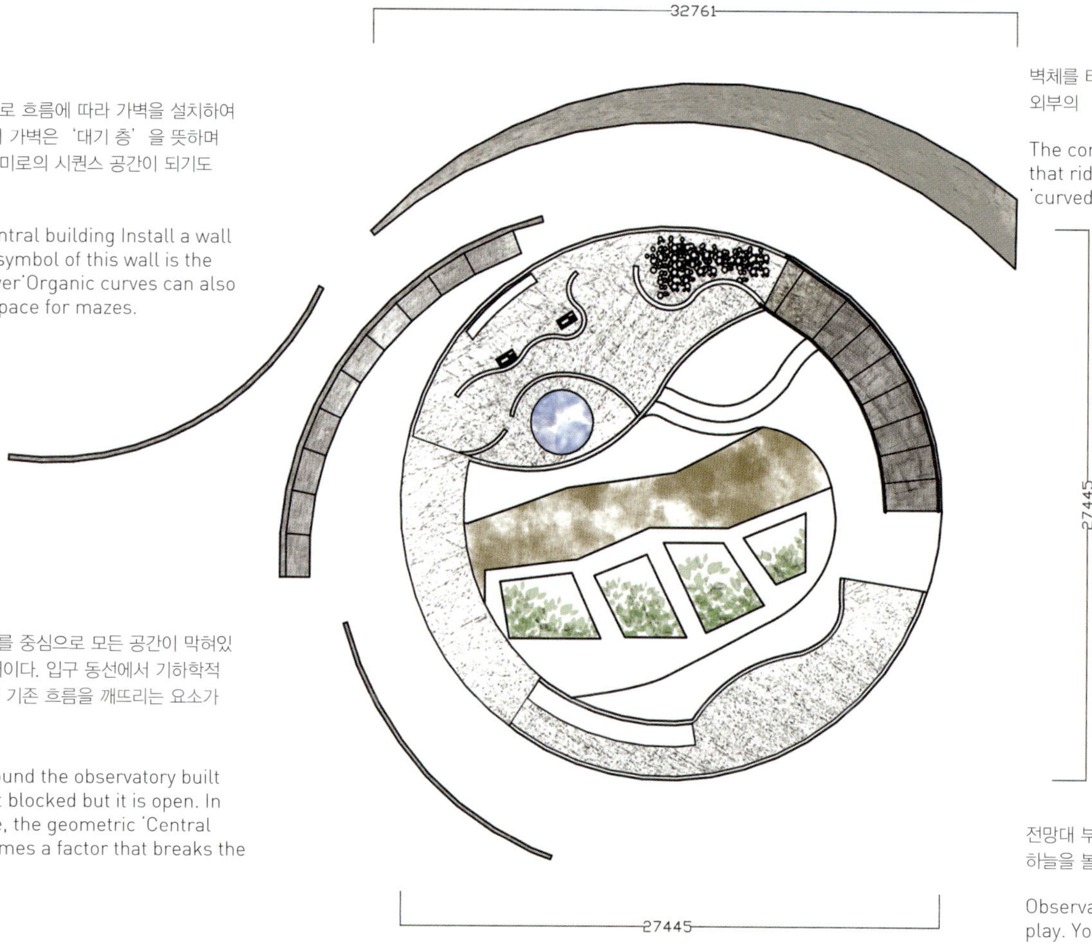

벽체를 타고 올라가는 '램프' 의 연결은 외부의 '곡선 벽체' 와 조화를 이룬다.

The connection of the 'ramp' that rides up the wall is the outer 'curved wall' Harmony.

옥상에 세운 전망대를 중심으로 모든 공간이 막혀있지 않고 개방된 형태이다. 입구 동선에서 기하학적 '중앙 공원 길' 은 기존 흐름을 깨뜨리는 요소가 된다.

All the space around the observatory built on the roof is not blocked but it is open. In the entrance line, the geometric 'Central Park Road' becomes a factor that breaks the existing flow.

전망대 부분은 판넬 전시와 함께 서울의 하늘을 볼 수있다.

Observation part with panel display. You can see the sky in Seoul.

ELEVATION

뚫린 곡선의 벽체는 카메라 프레임을 연상시켜 옥상 공원에서 서울의 전망을 한 눈에 담을 수 있다.

The wall of the curved curved wall reminds you of the camera frame, so you can see the view of Seoul from the rooftop park at a glance.

2nd FLOOR PLAN

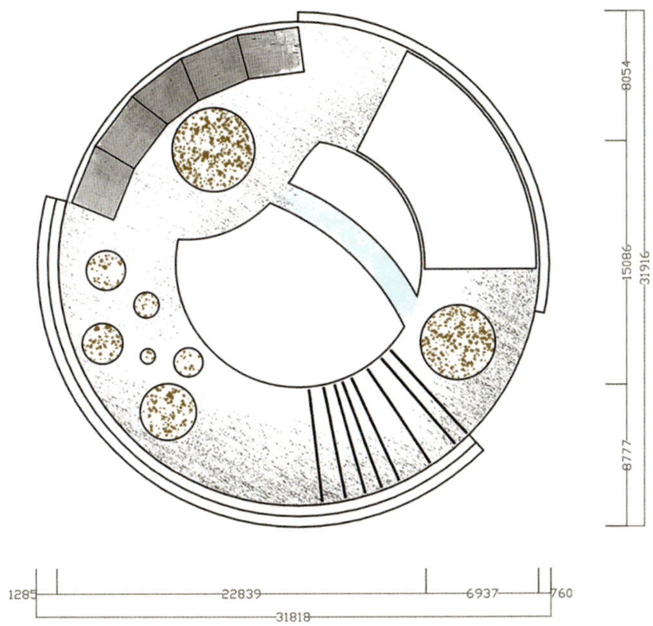

곡선의 흐름을 타고 올라온 곳엔 빛으로 이루어진 전망대가 펼쳐지는데, 이산화탄소량과 대기 중 오염도에 따라 색이 변한다. 또한 전망대 곳곳에 세워진 나무형태의 기둥들은 안개 속에 보이지 않는 나무의 실루엣을 표현하였다.

On the curved line, a light observatory is opened. The color changes according to the amount of carbon dioxide and the pollution level in the atmosphere. In addition, the wooden pillars standing on the observation deck represent the silhouette of the tree which is not seen in the mist.

2층의 중심 '공중 다리' 는 1층과의 호흡을 의미한다. 외부에서의 유기적 '곡선 벽체' 와 내부의 '램프계단' 이 대기 중에서 볼 수 없는 흐름을 구조적으로 표현한 것이다.

The central bridge on the second floor means a breath with the first floor. The organic 'curved wall' from the outside and the 'lamp stairs' inside are structurally representative of the invisible flow in the atmosphere.

다양한 색감의 파사드로 전망을 본 후에는 여러 겹으로 된 레이어 된 '유리 벽체' 가 있다.

After looking at the facade with various colors, there is a layered 'glass wall' with multiple layers.

CROSS SECTION

대기오염의 현상 중 피어오르는 효과를 오브제로 만들었다.

The object of the effect of the rising of the air pollution phenomenon was made.

DESIGN PROCESS

상징적인 의미를 가지는 건물의 파사드는 대기의 오염도에 따라 빛의 변화와, 안개분출 등 여러 효과가 나타난다. 그 중 1층의 전시공간에는 재난의 "현상-인식-극복" 순서의 공간들이 배치된다. 대기오염을 한눈에 볼 수 있는 디스플레이 전시, 그리고 공기의 답답함을 느낄 수 있도록 만들어진 조형물들은 우리에게 대기오염에 대한 문제를 보여준다.

The facade of a building that has a symbolic meaning has various effects such as a change of light and a mist generation depending on the pollution degree of the atmosphere. In the exhibition space on the first floor, space of order of "phenomenon - recognition - overcoming" of disaster is arranged. Display displays showing atmospheric pollution at a glance, and sculptures made to feel the air tightness show us the problem of air pollution.

1st FLOOR PLAN

옥상공간의 활용으로 개방감을 주기위해 2층은 전체적으로 커튼월을 사용하여 디자인하였다. 또한 공기의 팽창감을 표현하기 위하여 AIR를 이용한 벽채를 사용해 부풀어 오른 공간에서의 숨막힘을 느끼게 하였다.

The second floor was designed using a curtain wall as a whole to give a sense of openness through the use of the roof space. In addition, to express the feeling of air expansion, we used a wall bracket made of AIR to feel the breathing in the swollen space.

나무의 실루엣을 형상화한 기둥은 철로 이용하여 유기적 곡선형태를 디자인하였다. 가늘고 긴 형태의 철기둥이 얽혀서 하나의 형태를 이룬다.

The pillars that shape the silhouette of the trees are designed with organic curves using iron. The long and thin iron is entangled to form one.

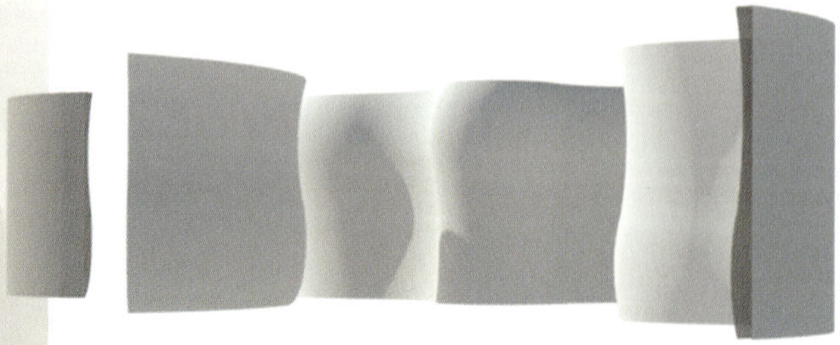

AXONOMETRIC

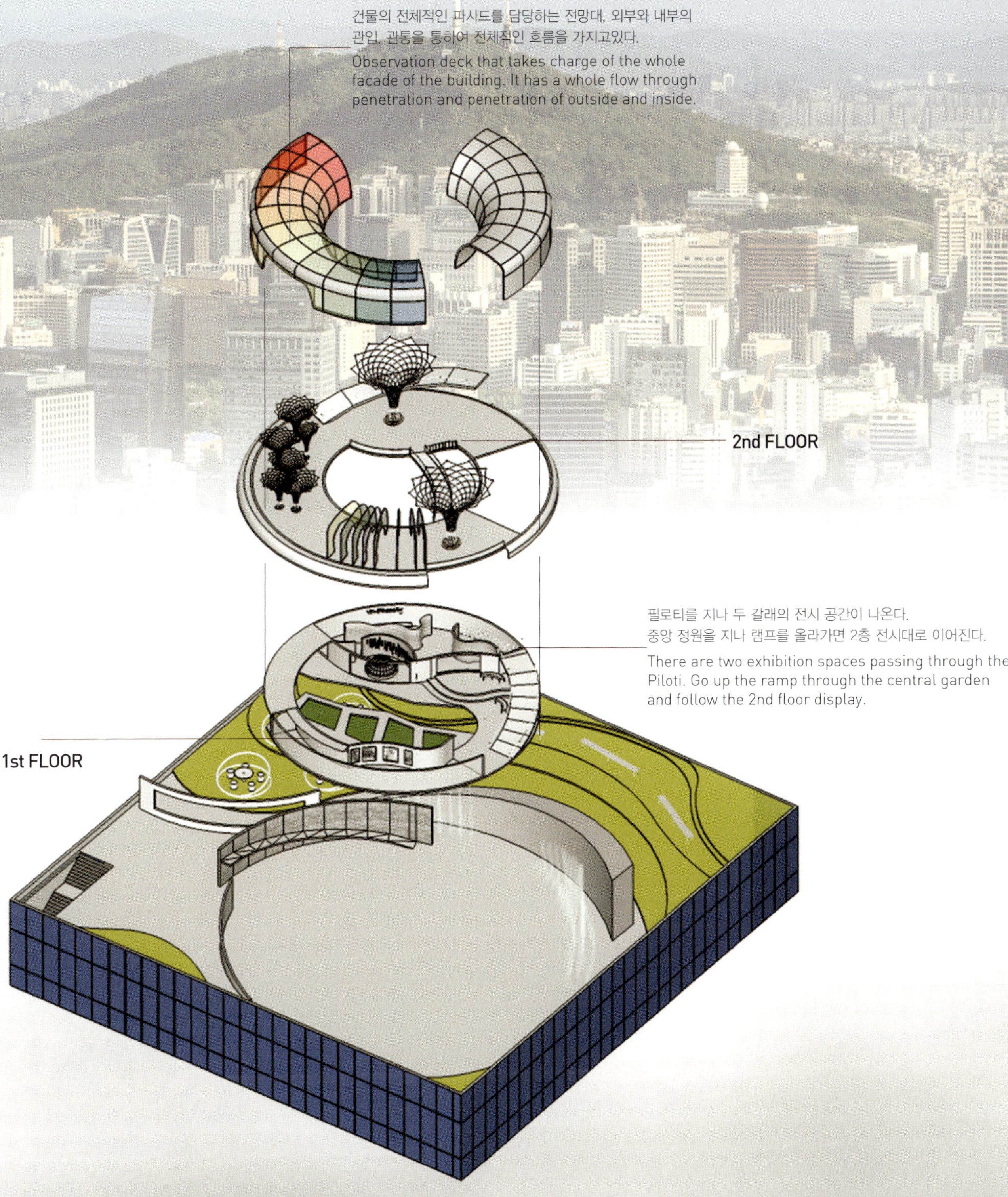

대기오염은 세계의 문제이며 모두가 관심을 가지고 변화하면 심각한 상황을 모면할 수 있다. 하지만 오염물질 배출의 원인은 광범위하기 때문에, 개인 그리고 국가가 같이 인식하고 관심을 가져야한다. 그렇기에 우리는 모두가 함께 지속적으로 관심을 가져야할 재난의 문제를 전시공간을 통해 사람들에게 알리고자 하였다. 맑은 날에는 수증기로, 안개가 많을 때는 밝은 빛을 이용함으로서, 다른 날에 대한 대비효과로 사람들이 이 재난에 대한 심각성을 인지하였으면 한다.

Air pollution is a problem of the world, and if everyone changes with care, serious situation can be saved. However, because the causes of pollutant emissions are wide, individuals and countries should be aware and interested. That is why we wanted to let people know about the disaster that everyone should continue to be interested in through the exhibition space. I hope that people will be aware of the seriousness of this disaster by using contrasting effects on other days by using water vapor on a clear day and bright light when there is a lot of fog.

1F SPACE PROGRAM

01_ 오염의 시작

옥상 정원을 가로질러 두 갈래길로 나누어지는 곳에는 대기오염에 대한 심각성을 체험할 수 있는 공간과 여러 판넬이 전시가 되어있는 전망대 전시관이 있다. 개방적인 전시공간은 답답함을 느낄 수 있는 체험공간과 대비된다.
이러한 개방형태는 환경오염에 대한 심각성을 모르는 우리들의 '무지함'을 표현하기 위함이고, 벽면에는 대기 오염을 재해석한 사진들과 대기오염에 대한 이야기가 이루어지는 판넬이 전시되어 있다.

01_ Start of pollution

There is a space where visitors can experience the seriousness of air pollution, and an observatory display panel where various panels are displayed at a place divided into two forks across the roof garden. The open exhibition space contrasts with the experience space where you can feel the stuffiness. This type of opening is to express our 'ignorance', which does not know the seriousness of environmental pollution. On the wall, there are photographs reinterpreting the air pollution and panels on the air pollution.

02_ 다가오는 재난

공간을 구성한 AIR로 만든 벽체는 유기적인 곡선 형태를 띄면서도 전시공간을 미로처럼 구성하고 있다. 또한 전시를 구경하러 온 사람이 공간속에서 답답함을 느끼며 전시를 관람하게 된다. 이 벽체를 따라 왼쪽으로 들어가면 첫 번째로 VR체험 공간이 나온다. 우리들의 생활습관으로 축적되어 발생한 스모그 현상으로 인해 황폐해진 우리의 세상을 비추는 영상이다.

02_ Upcoming Disaster

The wall made of AIR that composes the space has an organic curved shape, and the exhibition space is formed like a maze. Also, the person who came to see the exhibition feels the stuffiness in the space and watches the exhibition. If you go left along this wall, you will see the first VR experience space. It is a video that illuminates our world that has been ruined by the smog phenomenon that has been accumulated in our lifestyle.

03_ 보이지않는 세상

VR공간을 지나면 스모그로 가득 찬 전시공간을 볼 수 있다. 이 공간은 진동입자와 소리를 이용한 재난 체험 공간으로써 눈에 보이지 않는 대기 중 오염물질을 형상화하였고, 잘 보이지 않는 시야로 하여금 불안함, 공포, 두려움을 느낄 수 있다. 또한 시각이 불안정 하기때문에 그로 인하여 예민하게 반응하는 청각을 이용하였다. 촘촘하게 엉켜있는 입자들과 그 사이에서 뿜어져 나오는 스모그는 호기심을 자극하기도 하며 불쾌감을 느끼게 하기도 한다.

03_ Invisible World

After the VR space, you can see the exhibition space filled with smog. This space is a disaster experience space using vibration particles and sound, and it forms an invisible pollutant in the air, and an invisible sight can feel anxiety, fear, and fear. In addition, the visual acuity was unstable, so he used the sensitive hearing. The tightly entangled particles and the smog that emanates from them are both a source of curiosity and a discomfort.

세계의 대기오염
Air Pollution in the World

2F SPACE PROGRAM

01_ 그림자나무

블루스퀘어의 옥상이라는 SITE를 이용하여 서울을 한눈에 내려다 볼 수 있는 전망대를 디자인 하였다. 1F가 대기오염으로 인한 재난의 현상을 전시하였다면, 2F은 VIEW가 탁 트인 공간으로써 재난의 끝을 느끼게 한다. 또한 높게 솟아 있는 나무형태의 기둥들이 자리잡고 있는데 이것은 안개 속 안보이던 나무가 바람에 의해 서서히 실루엣을 나타내는 것을 형상화하였다.

01_ Shadow trees

The walls made of AIR, which compose the exhibition space, have an organic curved shape, while the exhibition space is shaped like a maze. Also, the person who came to see the exhibition feels the stuffiness in the space and watches the exhibition. If you go left along this wall, you will see the first VR experience space. It is a video that illuminates our world that has been ruined by the smog phenomenon that has been accumulated in our lifestyle.

02_ 파괴된 오존층

대기오염의 농도에 따라 빛이 변하는 전망대의 끝에는 벽체가 여러겹으로 LAYER된 공간이 나온다. 이것은 대기중 오염된 물질들로 파괴된 지구의 오존층을 표현하였다. 전망대의 곡선을 따라 LAYER된 공간은 다양한 높이감을 주며 공간의 변화를 가져온다.

02_ Destroyed ozone layer

At the end of the observatory, where the light changes according to the concentration of air pollution, there is a space where the walls are layered in multiple layers. This represented the earth's ozone layer that was destroyed by pollutants in the atmosphere. The space that is laid along the curve of the observatory gives a variety of height sense and changes the space.

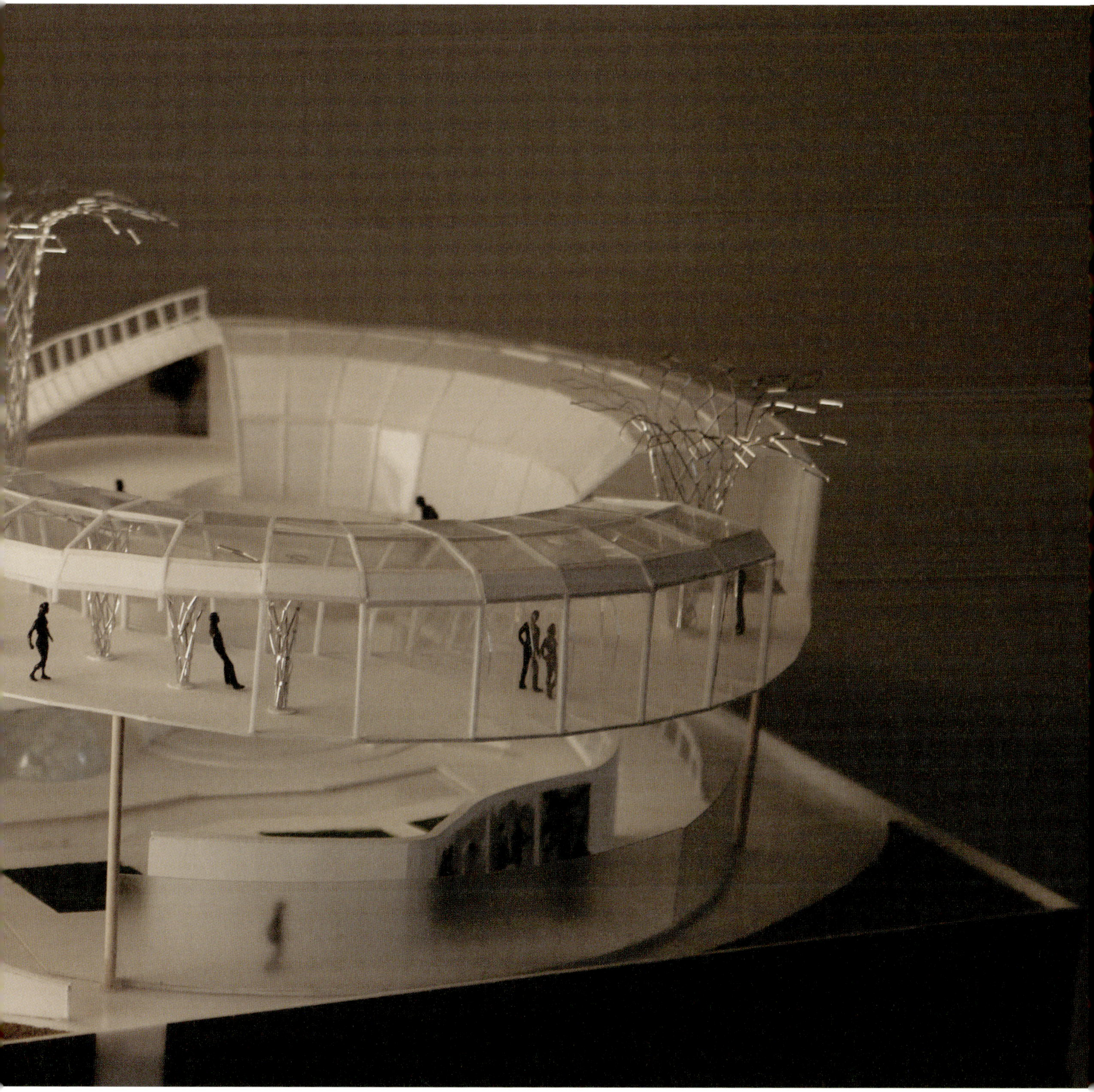

ATMOSPHERIC CLOUD

조하은 I Cho haeun 김단우 I Kim danu

STABILIZATION OF LIVELIHOOD

BACKGROUND

가설 : 평범한 낡은 다세대 주택이 불에 다 타버린 상황.
대부분 경제가 어려운 저소득층 사람들로 부상자를 제외한 이재민들은 당장 거주할 공간이 없는 상황.

경제적으로 어려운 사람들이 재난상황을 통해 집을 잃었을 때 과연 그들은 어떻게 생활의 안정을 찾을 수 있을까.
사회적으로 도움이 필요한 그들을 위하여 생활안정을 찾을 수 있도록 돕는 공간에는 어떤 것이 있을까.
현실적으로 당장 필요한 것은 그들이 머무를 수 있는 공간이다. 잠깐이나마 위로를 받고 안정을 찾아갈 수 있도록 그들을 위한 임시 거주 시설을 설계한다.

Hypothesis: Let's imagine there are old multi-family houses burned down.
Most of the people are receiving low-income and they are having difficulties with the economy, and the refugees except the injured have no house to live in right away. How could people find their livelihood in times of financial difficulty when they lose their homes due to disaster. What would it be there in the world that could help people find security for their socially needy lives?

PURPOSE

재난으로 피해 입은 저소득층 이재민들을 위한 임시거주시설을 계획하였다. 현재 저소득층을 위한 생활 지원 서비스는 갖추어져 있지만 많은 사람들이 혜택을 받기 어려운 현실이다. 때문에 저소득층 이재민들이 빠르게 생활의 안정을 찾을 수 있도록 일정 기간 동안 거주할 수 있는 임시거주공간을 계획하였다. 하지만 재난으로 인해 피해를 본 이재민이 거주하고 있지 않을 시에는 도움이 필요할 때 항상 달려가지만 정작 지쳤을 때 쉴 곳 없는 소방공무원 들을 위한 공간으로 활용되길 바랐다. 이재민들과 도움을 주는 소방관들이 서로 소통을 하며 재난이라는 상황 속 에너지를 얻어 안정을 찾는 공간이 될 수 있도록 설계하는 것이 우리의 목적이다.

What they need right now is the space where they can stay. To Design a temporary residence could be one way. With that, they can feel comfortable and safe for a short time. Moreover, we planned to make temporary shelters for low-income victims, who are suffering from the disaster. There are currently life support services for low-income people, but many people find it difficult to have the benefit. Thus, we have planned a temporary shelter for a period of time for the victims of low-income families to find their life stable. But, whenever the residence is vacant, we want it to be used as a space for firefighters. Therefore our goal is to design an infrastructure for victims and firefighters to communicate with one another in the place and find energy in the disaster and look for stability.

CONCEPT – FORT-DA

포르트 다는 실패를 멀리 던지고 끌어당기며 하는 놀이로 아가는 실패를 엄마의 상징으로 생각하고 엄마와 떨어져 있는 고통스러운 경험을 놀이로 승화하여 극복하려는 모습이라고 해석된다. 또한 이 놀이는 자신이 원하면 언제든지 대상이 돌아 올 수 있다는 능동적인 환상을 심어주기도 한다. 우리의 공간에서는 실패가 이재민들에게는 재난으로 인해 잃은 집을 상징한다. 또한 소방공무원들에겐 잠시 쉬었다 가는 놀이 공간으로써 활용된다.

FORT-DA is interpreted as a far-fetched, fading play. A baby thinks of failure as a symbol of her mother and maximizes her painful experiences. This play can also give a positive impression that the subject can come back anytime.

* FORT = 없다 DA = 있다

KEYWORD

1.Repetition

2.Stabilization

3.Overcome

4.Active

5.Play

6.spool

TARGET

재난으로 인해 집을 잃은 이재민과 소방관을 공간의 사용자로 설정한다.

In our space, failure symbolizes a home lost in disaster for the sufferers. It is used as a break and play area for fire department officials as well

SITE ANALYSIS

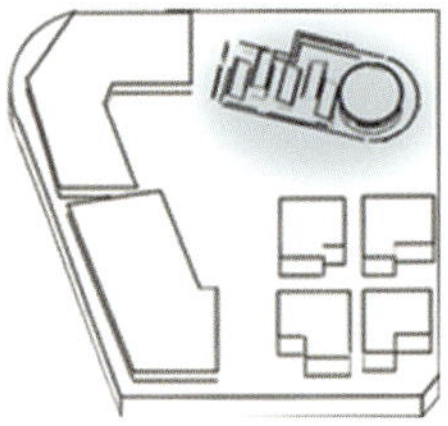

인천 광역시 소방서 및 남부 소방청

이 사이트는 소방 본부가 피해자와 가장 가까운 곳에서 그들을 돕기 위해 선정하였다. 또한 소방관은 소방서와 가까워 쉽게 접근하고 휴식 할 수 있다는 장점이 있다.

Incheon Nambu Fire Department fire station

If the temporary shelter is located close to the fire station, the shelter could be observed by the employees who are working in the fire stations and refugees can get help faster. We chose the location near fire stations because government employees could also easily access and relax

DESIGN PROCESS

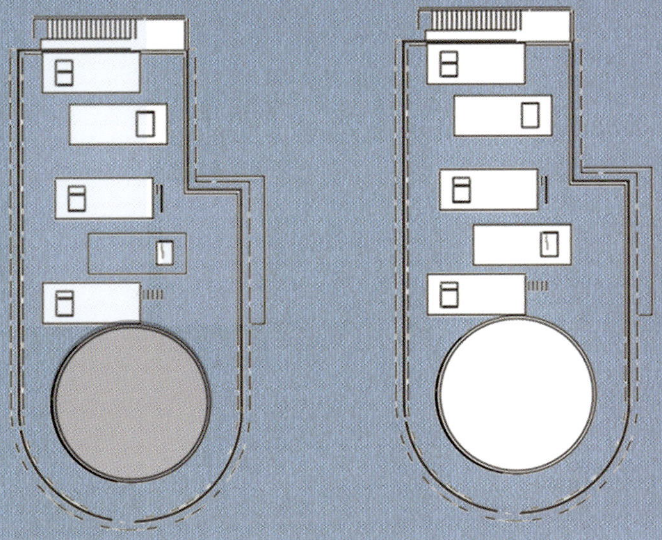

반복되는 행위로 인해 느끼는 안정감을 주는 포르트 다 놀이처럼 다양한 공간에 반복을 줌으로 반복적인 강박에서 오는 안정감과 공간의 통일성을 표현하고자 하였다.
사용자의 시선이 이동 할 때에 공간은 머물러 있지않고 계속해서 변화하게 된다.

I wanted to provide a more comfortable space for victims and firefighters by giving visual rhythm through repetition to the space like Fort Da Dao which gives a feeling of security due to repeated action.

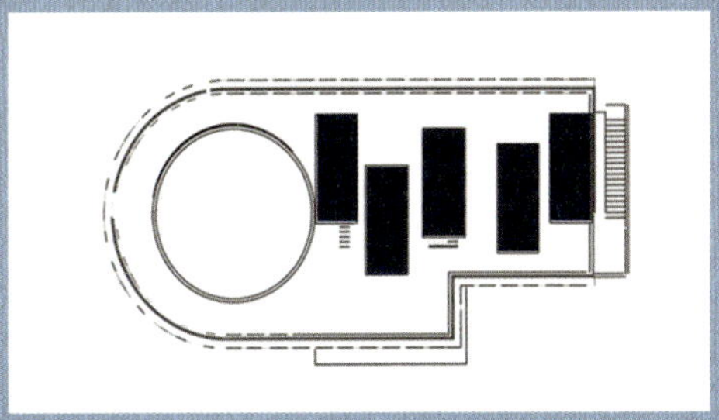

아치형태의 반복

아치 형태를 통해 시선이 이동할 때 반복을 통한 리듬감을 느끼게 된다. 포르트-다 놀이에서처럼 반복을 통해 느끼는 안정감을 시각적 리듬감으로 안정감을 표현하려 하였다. 사용자가 움직이면서 보였다 안보였다 하는 반복을 통해 오는 강박적 안정감을 표현하고자 하였다.

We would like to express the sense of stability and space that comes from repetitive movements by giving repetition to various spaces like the FORT-DA play, which gives us a sense of security. This can be relieved by our repeated action. When the user's gaze moves, space does not stay and changes continuously. If you move your eyes through the arch shape, you can feel the rhythm through repetition. As Fort-Da play, we wanted to express my sense of security through the visual rhythm of security that we felt through repetition.

Shadow

컨테이너박스

포르트 다는 실패를 멀리 던지고 끌어당기며 하는 놀이로 아가는 실패를 엄마의 상징으로 생각하고 엄마와 떨어져 있는 고통스러운 경험을 놀이로 승화하여 극복하려는 모습이라고 해석된다. 또한 이 놀이는 자신이 원하면 언제든지 대상이 돌아 올 수 있다는 능동적인 환상을 심어주기도 한다. 우리의 공간에서는 실패가 이재민들에게는 재난으로 인해 잃은 집을 상징한다. 또한 소방공무원들에겐 잠시 쉬었다 가는 놀이공간으로써 활용된다.

FORT-DA is interpreted as a far-fetched, fading play. A baby thinks of failure as a symbol of her mother and maximizes her painful experiences. This play can also give a positive impression that the subject can come back anytime.

Q.주택화재로 피해를 입은 저소득층 사람들을 위한 임시 거주 지원 시스템이 있나요?

A. 현재 주거피해를 입은 저소득층을 대상으로 2가구를 선정하여 피해를 입은 곳에 기업과 나라의 지원을 받아 재건축을 해주는 지원프로그램이 있지만 피해자들을 위한 임시거주시설은 따로 마련되어 있지 않습니다.

Q. Are there temporary housing support systems available for the low-income bracket who suffers from housing fires?

A.Currently there are support programs for low-income families. Governments select 2 families and those who are selected will receive reconstruction assistance in companies and countries that suffer from housing damage. But this is just for 2 households. To sum up, there are no temporary housing facilities for victims for now.

Q.저희가 그런 피해자들을 위한 공간으로 임시거주시설을 설계할 계획인데, 이 프로젝트에 대해 어떻게 생각하시나요?

A.현재 마련되어 있지 않은 대책이기 때문에 지어지기만 한다면 좋은 방향으로 잘 활용될 수 있을 거라고 생각합니다. 그러한 지원이 이루어진다면 피해자들을 위한 좋은 대책이 될 것 같습니다. 현재 저소득층 피해자를 위한 복구지원시스템이 있다하더라도 2가구밖에 지원을 해줄 수 없는 상황이고 그러한 대상자도 복구기간동안 거주할 공간이 마련되어 있는 상황이 아니기 때문에 그러한 곳이 지원이 된다면 피해자들을 위해 잘 활용 될 것 같습니다.

Q. We are planning to design and build a temporary residence as a place for such victims. What do you think about this project?

A. Well, because it is not prepared yet, I think it might be used well in a good direction if it is built. If such support is provided fully, it will be a good chance for victims. Even if there is a recovery supporting system, only two households will be supported. Because such people are not in situations, where there is no room for residence during the restoration period, it will be used well by the victim if such places will be supported for them.

Q.이재민들을 위해 이렇게 애쓰시는 소방공무원분들을 위해 그 시설에 소방공무원 분들을 위한 공간을 작게나마 함께 계획하고 싶은데 그 공간에 대해 어떻게 생각하시나요?

A.실질적으로 자주 활용할 수 있기에는 조금 어려움이 있겠지만 만약 그러한 공간이 함께 지어진다면 정말 기분 좋은 일인 것 같아요. 봉사를 하면서도 쉴 수 있는 공간이 있다면 더 큰 에너지가 생기니까요

Q. I'd like to make a small space together for fire department employees for those who work so hard at the facility. What do you think about my idea?

A. Actually, it's a little difficult to use it often. But it would be great if such a space would be given to the employees. If I would have a place to relax while serving, I will have more energy.

Q.좋은 답변들 감사합니다. 마지막으로 하고 싶은 말이 있으신가요?

A.화재 피해를 입은 저소득층 사람들은 사실상 생활의 안정을 찾기 힘든 경우가 많습니다. 그러나 그들을 위해 최선을 다해서 지원하고 있음에도 불구하고 부족한 부분이 많습니다. 그런데 이렇게 그들을 위한 공간을 생각한다는 것 자체가 저로써 너무 기쁘네요. 그런 공간이 지금은 없지만 여러 지원들을 통해 꼭 만들어졌으면 좋겠습니다.

Q. Good Answers. I really appreciate that. Do you have anything to say?

A. It is often difficult for low-income people to suffer from a fire disaster to find a stable life. But despite their best efforts to support them, there must be many drawbacks too. But I am so glad that at least there will be rooms for them. I hope various supports will be created.

PLAN&ZONING

1F

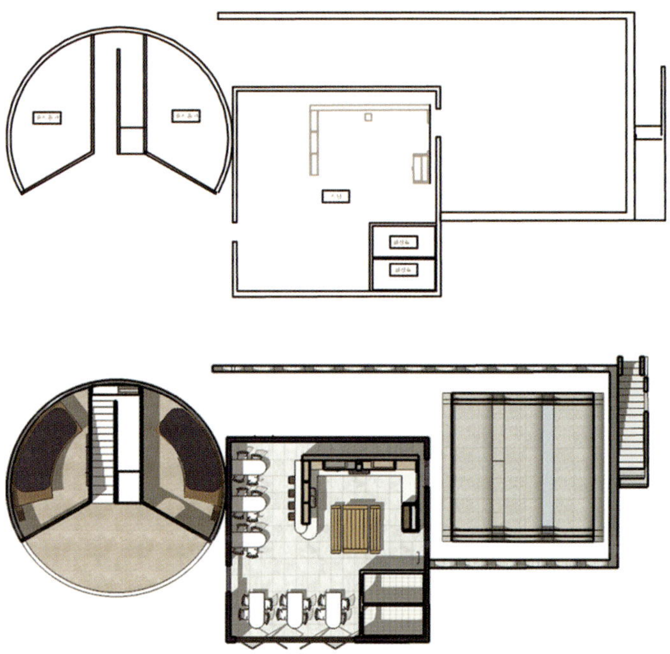

2F

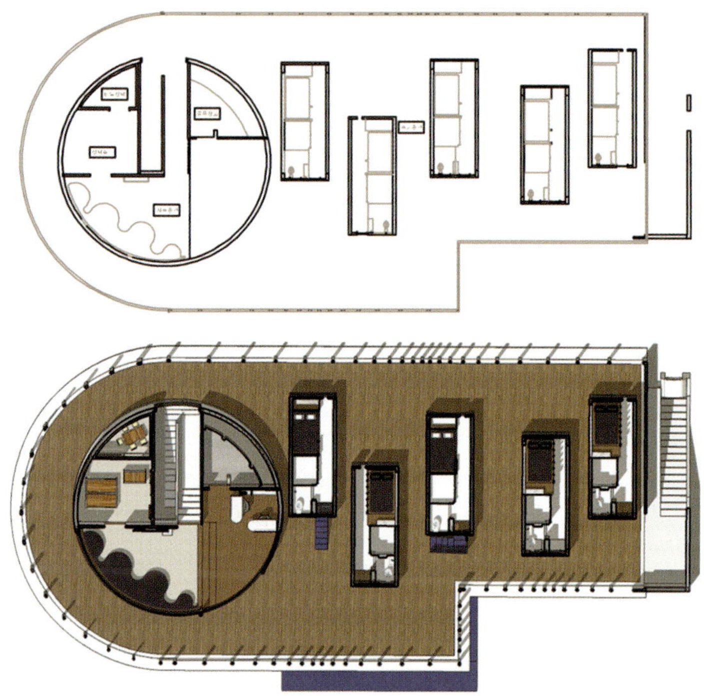

ISOMETRIC

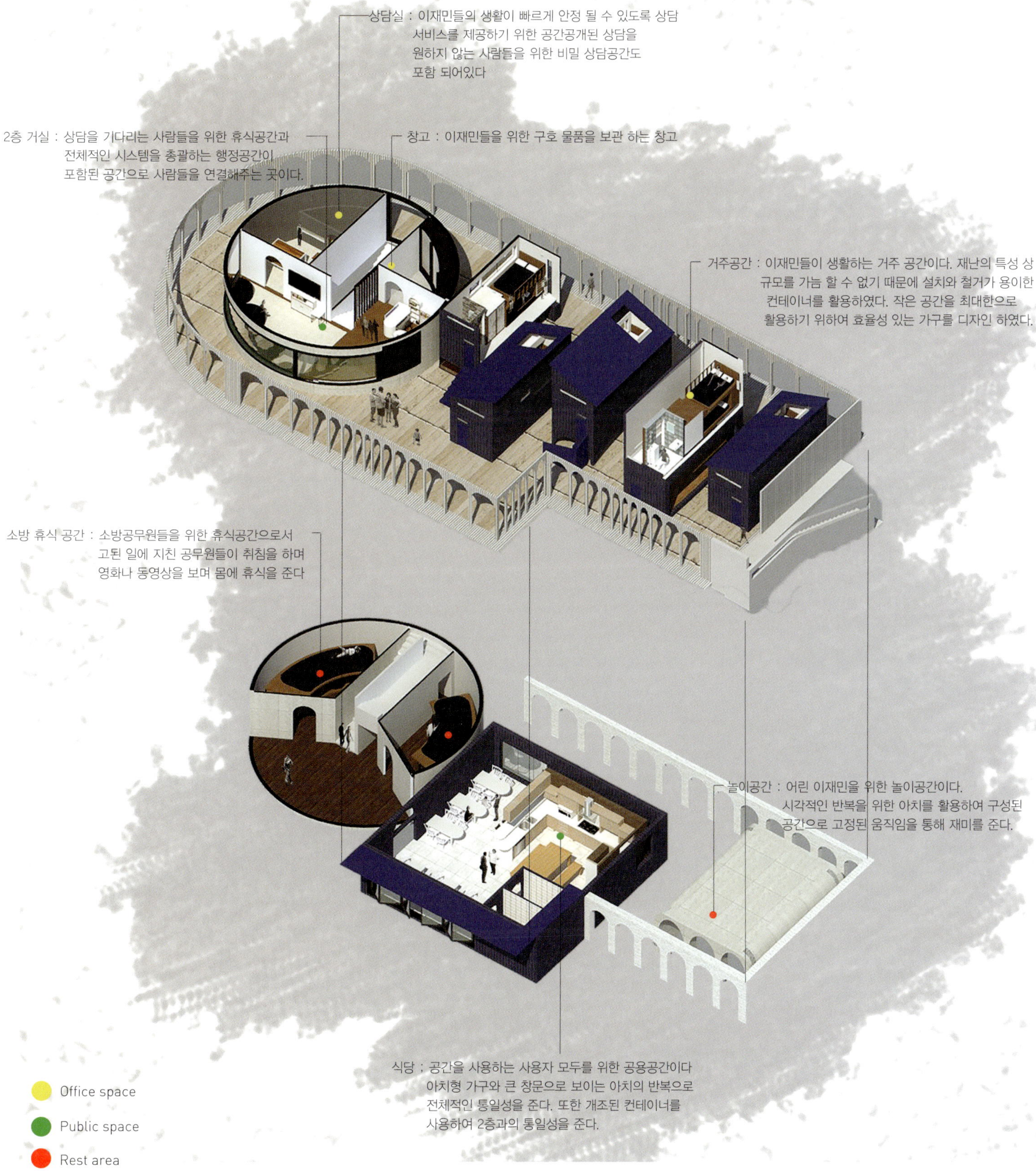
상담실 : 이재민들의 생활이 빠르게 안정 될 수 있도록 상담 서비스를 제공하기 위한 공간공개된 상담을 원하지 않는 사람들을 위한 비밀 상담공간도 포함 되어있다
2층 거실 : 상담을 기다리는 사람들을 위한 휴식공간과 전체적인 시스템을 총괄하는 행정공간이 포함된 공간으로 사람들을 연결해주는 곳이다.
창고 : 이재민들을 위한 구호 물품을 보관 하는 창고
거주공간 : 이재민들이 생활하는 거주 공간이다. 재난의 특성 상 규모를 가늠 할 수 없기 때문에 설치와 철거가 용이한 컨테이너를 활용하였다. 작은 공간을 최대한으로 활용하기 위하여 효율성 있는 가구를 디자인 하였다.
소방 휴식 공간 : 소방공무원들을 위한 휴식공간으로서 고된 일에 지친 공무원들이 취침을 하며 영화나 동영상을 보며 몸에 휴식을 준다
놀이공간 : 어린 이재민을 위한 놀이공간이다. 시각적인 반복을 위한 아치를 활용하여 구성된 공간으로 고정된 움직임을 통해 재미를 준다.
식당 : 공간을 사용하는 사용자 모두를 위한 공용공간이다 아치형 가구와 큰 창문으로 보이는 아치의 반복으로 전체적인 통일성을 준다. 또한 개조된 컨테이너를 사용하여 2층과의 통일성을 준다.
Office space
Public space
Rest area

상담 공간 : 이재민들을 위한 상담 공간이다. 상담공간과 비밀 상담 공간으로 나뉘어져있다.

컨테이너 식당 : 모든 사용자가이용하는 공용공간

2층 거실 : 공용공간 창고와 상담실, 1층과 2층을 이어주는 곳이다.

컨테이너 거주공간 : 이재민들이 생활 하는 거주 공간.

놀이공간 : 아치를 활용하여 시각적 움직임으로 놀이공간이다

소방관 휴식공간 : 소방관이 휴식하는 곳 누워서, 앉아서 영화나 영상을 볼 수 있다

조은비 l Jo Eunbi, 이현진 l Lee Hyunjin

Geometry disaster ground

기하학 재난 놀이터

Background

북극 곰의 삶의 터전인 빙하가 지구온난화로 인해 점점 녹아 들고있다. 인간의 삶이 풍족해질수록 북극에 있는 동물들의 삶은 피폐해지고 있다. 지구는 과연 인간만의 것인가? 지오메트리 디재스터 그라운드(지구온난화 재난 체험 놀이터)를 통해 북극곰의 눈물을 체험한다.

Because of global warming, glaciers, the home of polar bears, are melting constantly. As human life grows richer and richer, the lives of animals in the Arctic are becoming more impoverished. Is the earth only for people? Now, we experience the tears of polar bears through the global warming disaster experiencing area.

Ploblem

지구온난화란 지구의 평균 기온이 점점 높아지는 현상으로, 산업 혁명 이후 인구 증가와 산업화에 따라 화석연료의 사용이 늘어나 온실 가스 배출량이 증가하고, 무분별한 삼림 벌채로 대기 중의 온실 가스농도가 높아지면서 지구의 평균 기온이 상승 하는 지구 온난화 현상이 나타나고 있다. 이로 인해 극지방과 고산 지역의 빙하가 녹아 평균 해수면이 상승하여 해안 저지대가 침수 될 위험해 처했다.

Global warming is a phenomenon that causes the average temperature of the earth to rise after the industrial revolution. As the population increases and the use of fossil fuels increase with industrialization, greenhouse gas emissions are rising, and the average greenhouse gas emissions are right before the deforestation. As a result, glaciers in the polar regions and alpine regions have melted and the average sea level has risen. Therefore, the coastal lowlands are under the threat of flood.

Solution plan

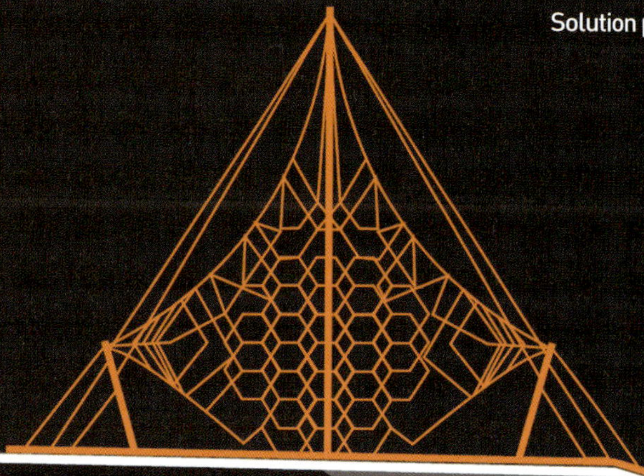

경각심 Awakening & 일침 Advice

아직은 당장 우리에게 보이지 않는 것들로 하여 체험하는 공간을 구성한다. 미래에 지구를 이끌어갈 이들에게 좀 더 친근하게 다가가기 위해 놀이적으로 접근한다.

We are now making a room for the things that we can not see at this moment. We playfully approach the earth to become more familiar with future Earth leaders.

Site analysis

서울 영등포구 선유로 343 (선유도공원)

선유도 공원은 옛 정수장을 활용한 국내 최초 재활용 생태공원으로 물을 정화시키는 공간이었다. 우리는 이 Geometry disaster ground를 이 공간에 둠으로 지구를 정화시킨다는 의미를 두었고, 이에 지구온난화라는 재난을 담을 수 있는 공간이라 생각하게 되어 선정하였다. 선유도 공원에 [2017 대한민국 재난 전] 이라는 재난 놀이터를 연다.

Seonyudo Park -343 Seonyu street yeondengpo-gu, Seoul, South Korea.

Sunyudo Park is South Korea's first recycled ecological park and was used to clean the water. We decided to emphasize purifying the earth by placing this global warming-disaster experiencing area in this space and this is why we chose the disaster area of global warming. We will open a disaster playground called [Before 2017 Korea Disaster] in Seonyudo Park.

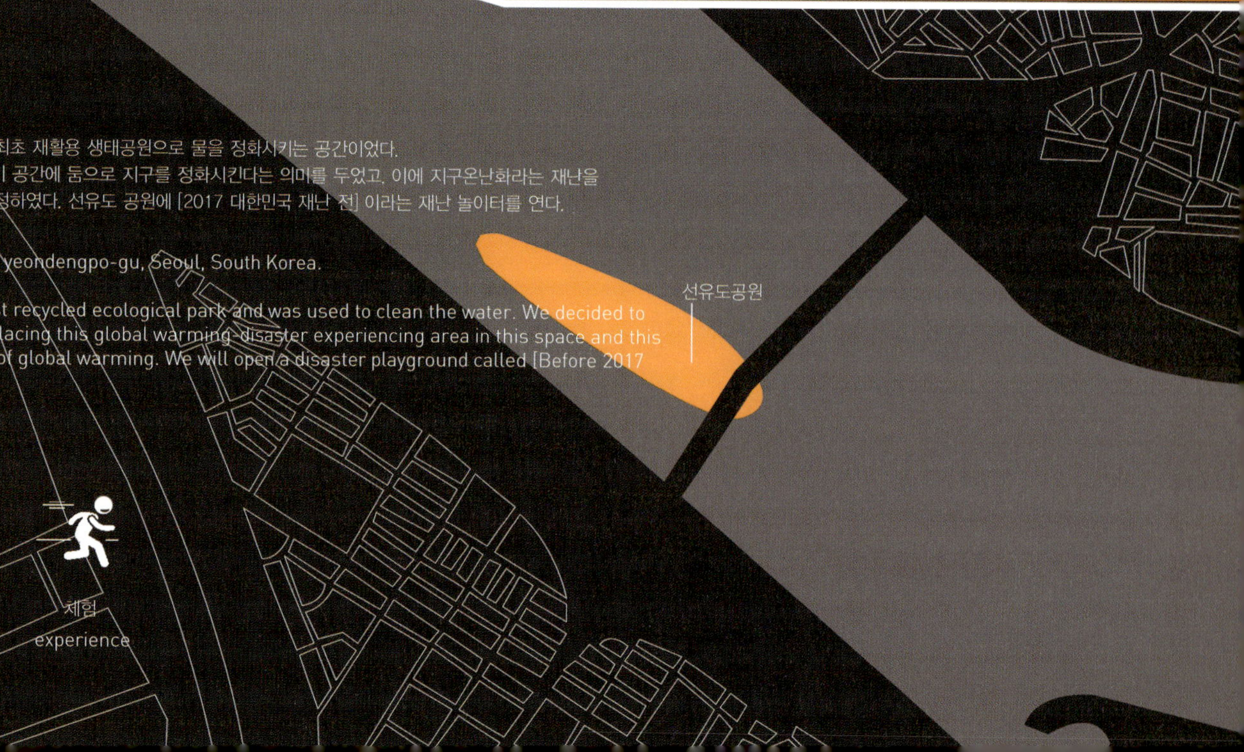

접근성 accessibility

연관성 correlation

체험 experience

Concept

지구온난화는 지구 표면의 평균온도가 상승하는 현상이다. 자연그대로가 아닌 사회의 발전으로 인한 자연을 거스르는 인류의 활동을 업보로 생각하여 자연의 유기적인 형태에 반대되는 기하학적 표현을 주로 사용한 구성주의작품 중 'Gometric Composition' 을 컨셉으로 하였다. 순수 형태를 구성하고 있는 구성주의를 모티브로 하여 기하학적인 형태들을 사용함으로써 현재상황을 꾸밈없이 보여주려고 한다.

Global warming is a phenomenon in which the average temperature of the earth's surface rises. We believe that human activities that go against nature, not just in nature, are karma. Thus, we are using Russian constructivism, which is a geometric statement that is mostly opposed to the organic form of nature. We would like to show the current situation honestly by using a simple geometry of structuralism in its pure form.

기하학적인 + 재난 + 놀이터 = ?

Geometric + Disaster + Playground = ?

Mass study

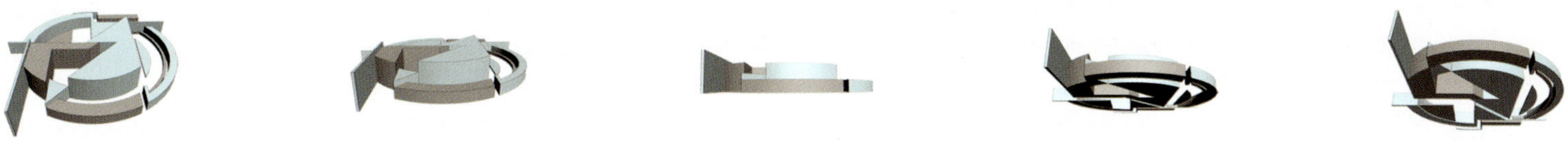

구성주의 작품 중 'Geometric Composition' 에서 볼 수 있는 큰 덩어리의 매스를 수직으로 올렸다.

We put up a large mass vertically, which can be seen in the composition of Geometric Composition.

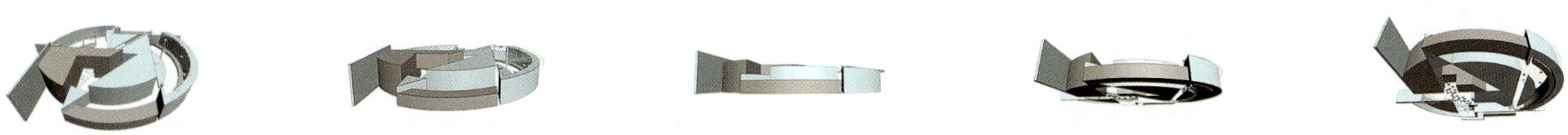

단조로웠던 매스에 재난이라는 주제에 맞는 공간을 연출하기위해 매스를 더하고 올린다.

We inserted the mass to create the place that suits our theme in an ordinary mass.

올리기, 비우기를 하며 공간에 볼륨감을 넣고 재난놀이터라는 주제에 맞는 매스를 형성한다.

We then form a mass that fits our concept by putting massing in the place with lifting and emptying it.

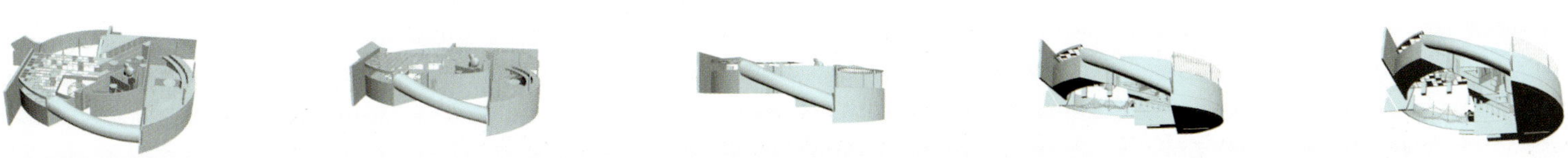

재난놀이터라는 공간에 좀 더 어울리는 공간을 형성한다. 좀 더 재밌는 놀이터와 좀 더 위험한 재난상황을 느낄 수 있는 공간을 연출한다.

We create spaces that create more suitable space for disaster playground space and that can feel a dangerous disaster situation.

Isometric

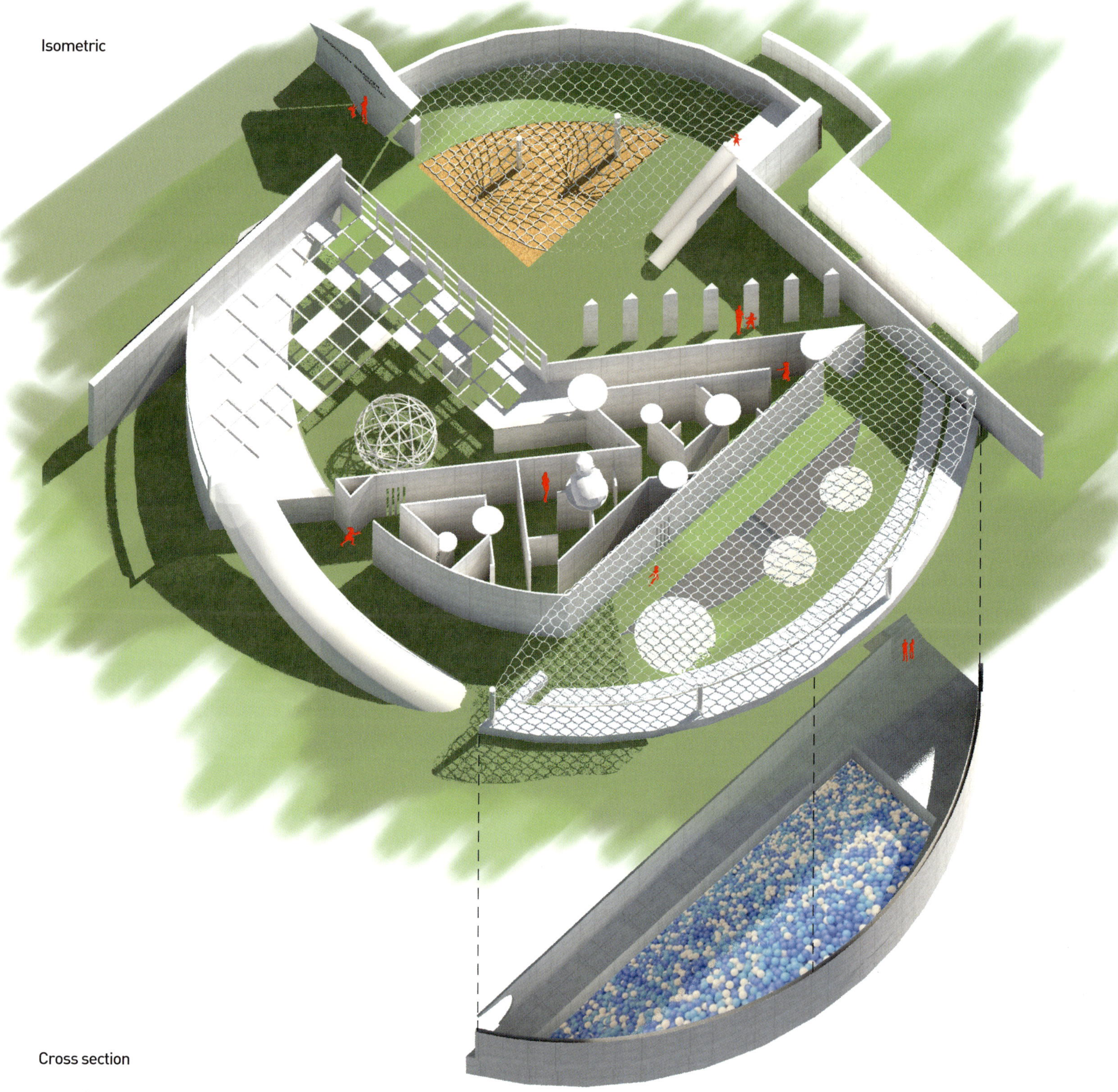

Cross section

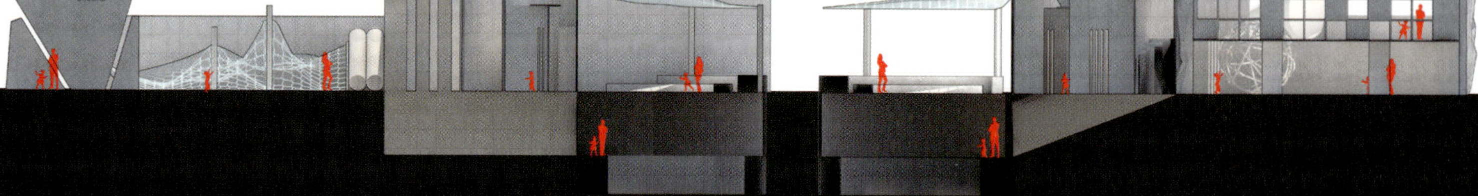

Direction space

메스의 데드-스페이스인 부분을 빙하가 갈라지고 있는 느낌으로 분리시켜 연출

Create the dead space of the mass by separating, feeling like the glaciers are breaking.

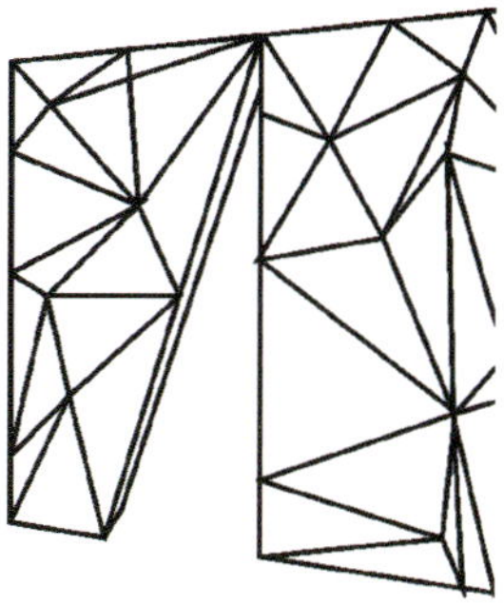

오존층이 파괴된 지구를 상징하는 구조물로 지구온난화가 가지고 오는 혼란스러움을 나타냄

It shows confusion caused by global warming. with a structure that represents the destruction of the earth.

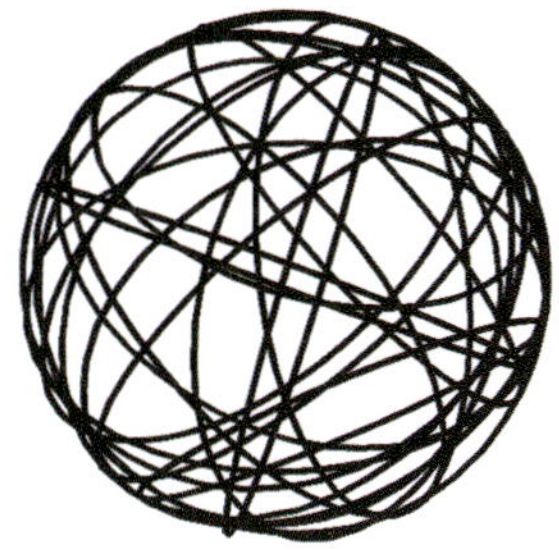

미로 크레바스위에 올라 먼 산을 바라보는 북극곰의 모습을 연출

Creating a symbolic icon- a polar bear looking over a distant mountain on a maza crevasse.

Zoning

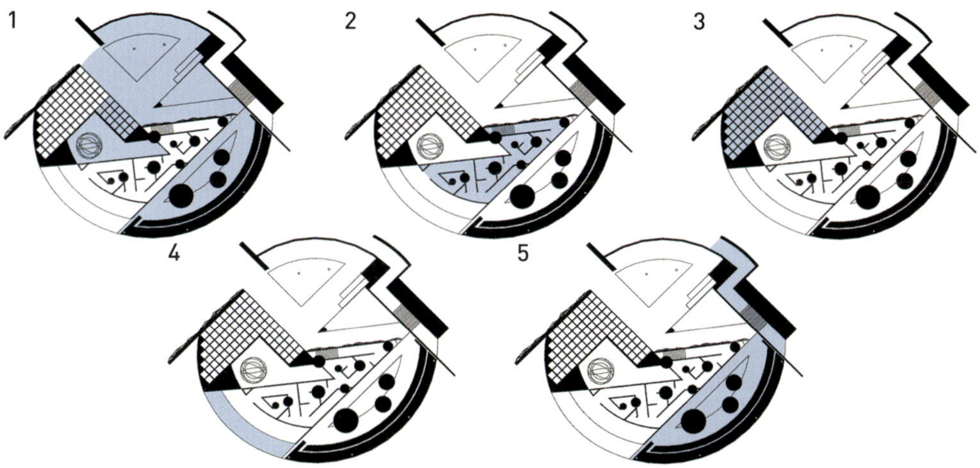

1. 위험한 놀이터 컨셉의 구조물과 상징적 오브제 설치
2. 크레바스를 모티브로 한 미로 설치
3. 녹아버린 빙하조각을 다리로 표현
4. 지상에서 지하로 내려가는 미끄럼틀
5. 수면의 잠김을 표현한 볼풀장

1. Install a structure and a symbolic objet under the concept of a dangerous playground
2. Install a maze according to the motif of crevasse
3. expression pieces of broken and melted glaciers as a bridge
4. A slide that goes underground from the ground floor.
5. A ball pool which expressed the immersion of the sea

Icon

방관 지구온난화에 대해 무심한 현재 상황에 대한 심각성을 느끼지 못하고 있는 현재 방문자들은 해수를 그물로 표현된 기구를 체험한다. 가벼운 놀이 뒤에 빙하가 점점 사라지고 있어 밟고 서 있을 곳이 없어지는 현실을 손과 발의 자유롭지 못함을 통해 빙하가 녹아 물속으로 떨어지는 체험으로 보여준다.

Observing Current visitors who do not take it serious about current situation about global warming will experience apparatus expressed with a net of sea water. After the play, the glacier gradually disappears and it shows an effect that from the standing position, glacier melts and the hands and feet are frozen and fall into the water. To make it clear, the visitors can feel that they have nowhere to stay anymore by looking at the disappearing glaciers.

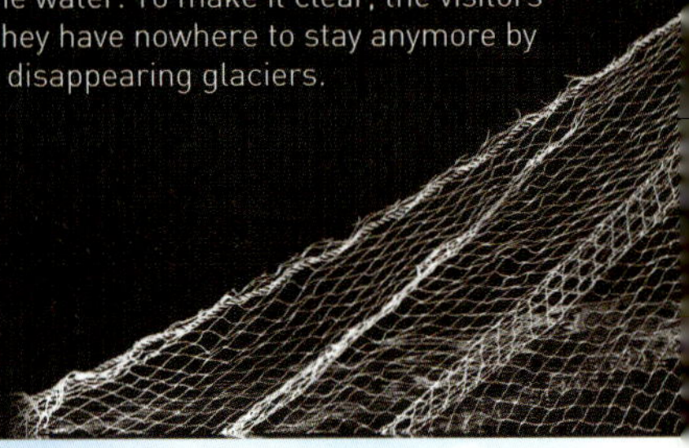

긴장

넓었던 공간이 벽에 의해 줄어들면서 밟을 수 있는 대지 또한 점점 줄어든다. 대지가 줄어들면서 가볍게 생각했던 상황이 심각해짐을 인지한다. 긴장감을 조성하게 하고, 길 끝에 뭐가 있을지 궁금증을 유발시킨다.

Nervousness

As the wide space is reduced by the walls, the walking area is also getting narrow. As the land shrinks, visitors aware the serious situation. It makes you nervous and raises the curiosity of what is at the end of the way.

불안

앞에 놓인 해수그물이 천장으로 올라가 있어 지구 온난화가 진행되어 해수가 상승했음을 표현한다.물속에 잠겨있음을 표현한 공간으로 그물을 통해 생긴 그림자는 그 느낌을 더한다. 물속에서 몸을 쉽게 제어할 수 없음을 놀이기구를 이용하여 체험하고, 시각적으로 보이드한 공간을 연출하여 이전 공간과 연계되어 극적인 효과를 주는 동시에 바닥의 유리를 통해 지하와 시각적으로 공유되어 불안한 감정을 느낀다.

Anxiety

The Seawater nets rise up to the ceiling as global warming progresses. Shadows created through the net make people feel like that they are going to be submerged in water. Experiencing the difficulty of moving in the water through the equipment. It visually creates space that has a dramatic effect in relation to the previous space. Furthermore, visitors feel anxiety by visually sharing with the underground space through the floor glass at the same time.

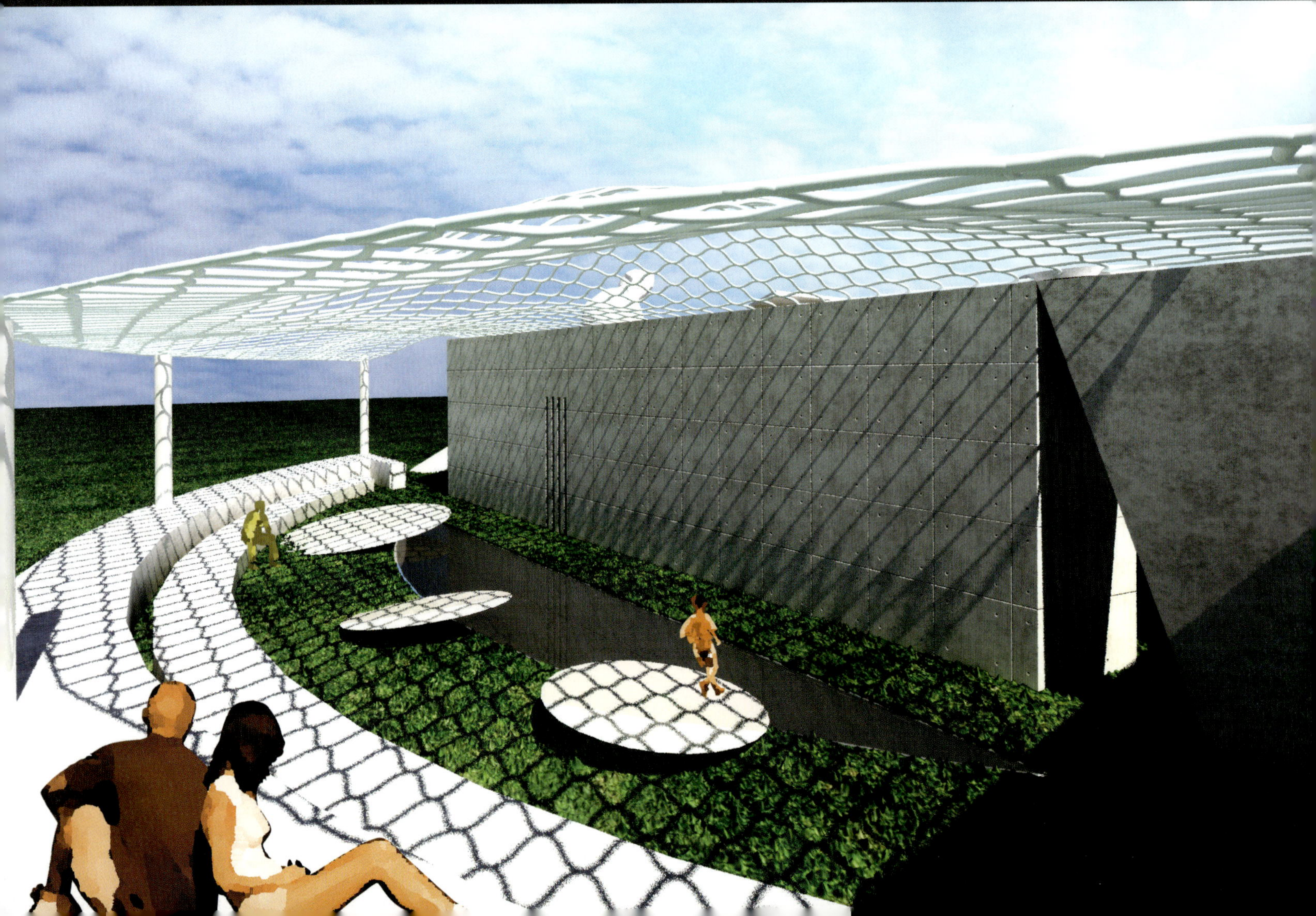

사색

균열에 의해 생긴 크레바스를 모티브로 한 미로의 중간 쯤에서 먼 산을 바라보고 있는 북극곰을 만나게 된다. 현재 일어나는 지구온난화의 상황들, 다가올 미래에 대한 사색을 하는 공간으로 이를 보면서 지구온난화의 심각성을 스스로 생각할 수 있는 사색공간이다. 벽에 뚫린 슬리트에 의해 생기는 빛은 시각적 변화를 수반하여 공간을 이화시킨다.

Contemplation

Visitors will meet a polar bear facing a distant mountain from the middle of a maze with the theme of the crack created by the crevasse. This is the place where people can think deeply about the seriousness of the current situations and even the situations yet to come. The light from the slit can alter the space to accompany visual changes.

공허 사색하고 난 뒤 감정은 공허하다. 빙산이 녹아 작게 부서진 해빙을 투명한 유리와 불투명한 도형을 이용하여 표현하여 막혀있는 발판을 찾아 이동하며 좀 전에 지나온 북극곰과 연계되어 북극곰의 비애를 느낄 수 있다.

Emptiness After the consideration of those situations, we feel sad. As we move around, we can find out polar bear's sorrow, using transparent glass and opaque figures to express a broken sea ice due to melting.

성찰

미끄럼틀이 있어 중력에 의해 떨어져 지하로 이동한다. 천장에 뚫린 천창에 의해 지나온 세 번째 공간과 시각적으로 공유되어 창을 통해 보이는 해수 그물을 바라보며 해수가 한층 더 상승했음을 인지한다. 물속에 완전하게 잠긴 것으로 볼 풀장을 헤쳐나오며 지난날의 과거와 앞으로 일어날 일들을 성찰한다.

Contemplation

There is a slide, which is dropped by gravity and moves to the underground. The third space passed by the light on the ceiling shared visually, recognizing that sea water has risen further by looking at the seawater net that they can see through the window. Considering that they are totally submerged in the water, people go through the ball pool. And those activities will give them time to reflect themselves the problem of the global warming once more.

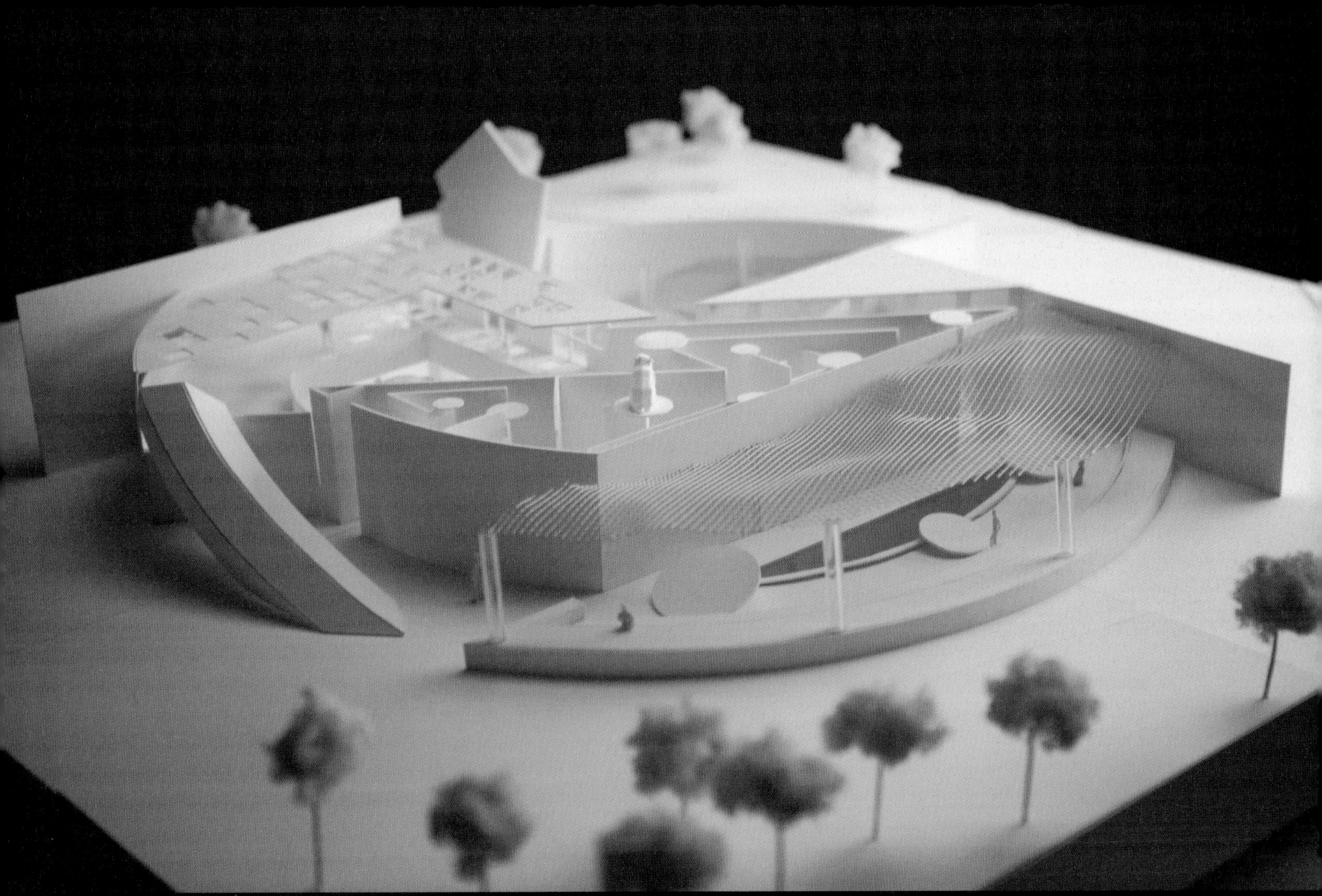

2018
INTERNATIONAL DESIGN EXCHANGE PROJECT

The 9th Hidden Space Project

RECUSED; Space & People

The Cass London Metropolitan University
The Sir John Cass Faculty of Art, Architecture and Design
Interior Design BA (Hons)
UK

Dongyang Mirae University
Dept. of Interior Environmental Design, in School of Living and Environmental Engineering
Rep. of KOREA

Pétur Thomsen, <Imported Landscape>

The 9th Hidden Space Project
RECUSED; Space & People

19 June – 4 July 2010
HUN
IAS
IDN
Intl Architectural Showcase
Presented by foreign embassies and cultural institutes in collaboration with The British Council and The Architecture Foundation for the London Festival of Architecture 2010

JPN
KOR
LVA
Living Bridges
A UK and Korea collaboration
26. Korean Cultural Centre and London Metropolitan University
An exhibition of work by 25 students from three collaborative universities; London Metropolitan, Kyung Won and Dong Yang. The project brief focused on urban and community issues shared by the UK and Korea and asked the students to create a 'living bridge' structure that is both an architectural metaphor for community support and a space for the communities to enjoy, integrate and communicate. Each student has built a model and will show the structure, its location and context.
The work from this project has been compiled in a book that will be published alongside the exhibition. Living Bridges has given us a valuable insight into how young creative minds work, and how they deal with physical evidence and community challenges.
31
16 Goulston Street, London E1 7TP
22 June– 27 June
10am– 5pm
Private View: 24 June, 6–8pm

2010
INTERNATIONAL DESIGN EXCHANGE PROJECT

The 1st Hidden Space Project

LIVING BRIDGES

The Cass London Metropolitan University
Dongyang Mirae University
Kyungwon University

Symposium Hidden Spaces :
Wednesday 27th June 10.30am-3.30pm
Gensler Architects
Aldgate House
33 Aldgate High Street
London EC3N 1AH

2012
INTERNATIONAL DESIGN EXCHANGE PROJECT

The 3rd Hidden Space Project

HIGH STREET

The Cass London Metropolitan University
Dongyang Mirae University
Kyungwon University

LONDON
metropolitan
university

Gerard Byrne: 1984 and Beyond

2014

INTERNATIONAL DESIGN EXCHANGE PROJECT

The 5th Hidden Space Project

THE RE-ENERGISING SPACE

The Cass London Metropolitan University
Dongyang Mirae University
Hunan University

FHHS2010 - *The 1st Hidden Space Project* "LIVING BRIDGES"

FHHS2011 - *The 2nd Hidden Space Project* "URBAN ROOFTOPS"

FHHS2012 - *The 3rd Hidden Space Project* "HIGH STREET"

FHHS2013 - *The 4th Hidden Space Project* "THE RISE OF THE CREATIVE DISTRICT"

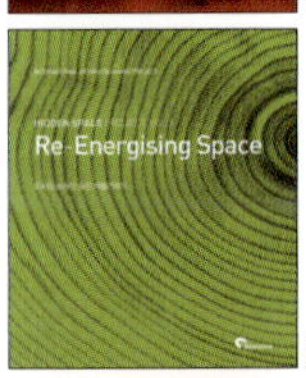

FHHS2014 - *The 5th Hidden Space Project* "RE-ENERGISING SPACE"

FHHS2015 - *The 6th Hidden Space Project* "RE-EXPERIENCE SPACE"

FHHS2016 - *The 7th Hidden Space Project* "SPATIAL HAPPINESS"

FHHS2017 - *The 8th Hidden Space Project* "REINTERPRETATION"

FHHS2018 - *The 9th Hidden Space Project* "RECUSED: Space & People"

* FHHS (Future Horizon for Hidden Space)

2010년 [LIVING BRIDGES]주제로 영국 런던 메트로폴리탄대학과 한국의 가천대학교, 한국의 동양미래대학교가 공동으로 런던 아키텍 페스티벌 참가하면서 시작된 FHHS는 우리시대 숨겨진 공간들을 10년간 10개의 주제로 교류전을 기획, 현재 9회까지 진행했으며 이후 주제를 현재 기획중이다.
FHHS는 이후 계속해서 우리주변의 숨겨진 공간들을 찾아 열정과 지혜, 그리고 성찰의 노정(路程) 을 계속해 나갈 예정이다.

2018

INTERNATIONAL DESIGN EXCHANGE PROJECT

The 9th Hidden Space Proect

RECUSED; Space & People

Published By 2018 RECUSED; Space & People Project
(Kaye Newman / Janette Harris / Park Youngtae)

Total Design
Park Youngtae | DONGYANG MIRAE University
만든이 | 동양미래대학교 실내환경디자인과 박영태
편집, 기획, 표지 디자인 | 박영태 / 편집보조 | 윤성준, 조명희

펴낸날 | 2018. 8. 30. 1쇄 발행
발행인 | 김기현
발행처 | spacetime · 시공문화사
등 록 | 1993년 3월12일 1-1493
주 소 | 120-796 서울시 서대문구 독립문공원길 13 극동프라자 5층
전 화 | 02-3147-1212
팩 스 | 02-3147-2626
ISBN | 978-89-5592-404-6

정가 | 18,000원